THE CYBER ATTACK
SURVIVAL MANUAL

NICK SELBY · HEATHER VESCENT

THE **CYBER ATTACK SURVIVAL MANUAL**

TOOLS FOR SURVIVING EVERYTHING FROM IDENTITY THEFT TO THE DIGITAL APOCALYPSE

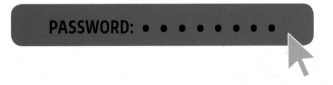

PASSWORD: • • • • • • • •

Illustrations by
ERIC CHOW
and Conor Buckley

weldon**owen**

CONTENTS

HACK YOUR LIFE

HACK SOCIETY

HACK THE WORLD

THIS BOOK IS ALREADY OUT OF DATE
(AND THAT'S OKAY)

Every day there seems to be a new story about cybercrime: millions of credit cards stolen, private celebrity photos leaked, foreign agents interfering at the highest levels of government. It's hard for even the best-informed reader to know how much of this is real versus scare-mongering clickbait, and how to react regardless. Sadly, many people either become so paralyzed by fear that they vacillate between different strategies for too long/ Conversely, some decide it's all too much and try to ignore the topic completely.

The thing is, each of us is utterly reliant on cybersecurity in ways both obvious and unexpected. As a police officer, I've brusquely knocked on the door of the suspect in a cyber case only to find a 78-year-old retiree, innocent of anything but a yen for some specialized, icky, but legal pornography. Our man made a rookie mistake by going for the free icky-but-legal porn, unaware of the first rule of the web: if you can't figure out how someone makes money on a site, you're the product. Criminals had planted malware in his naughty movies and were renting cyber scammers remote access to his computer, unbeknownst to him.

Some of the hacks we describe will be old news tomorrow. Some will take on new and more insidious forms. And something new will pop up every time you turn around. That's okay—this book gives you the tools you need to understand what your digital footprint looks like to criminals, advertisers, investigators, and governments, and how to figure out and fix your vulnerabilities even as the specific threats change.

We can't tell you everything that might happen to you—some of next week's threats are being cooked up right now in basements and labs from Missouri to Moldova. But we can tell you how to reduce your risks no matter what. Security experts like to talk about OPSEC (operational security). And OPSEC is OPSEC—today and forever. It's not about specific dangers, it's about a mind-set of preparedness.

Understanding your digital universe and the consequences of your actions will reduce the things that can make you a victim, without your having to miss out everything the internet has to offer. This book will help you better understand the kinds of threats out there, and give you the tools and perspective to protect yourself. The rest is up to you.

NICK SELBY

ALSO, THIS BOOK WILL FREAK YOU OUT

(AND THAT'S ALSO OKAY)

To put it simply, you're in danger. Your identity, your bank accounts, your kids, and even your government are vulnerable to attack from cyber criminals around the world. That should freak you out. But this book is much more than a collection scary stories (although it's that too). It's also a toolkit for protecting yourself and your data in an increasingly dangerous online world.

The digital age has given us a dazzling array of products and services at our fingertips, but also created new and often unexpected problems. Security technology will continue to get better—and criminals will keep finding new ways to get around that technology. That's where we come in.

How to get your head around security in the modern age? Most people want to know first and foremost how to avoid getting hacked. That's the wrong mindset. It's almost inevitable that you're going to be hacked at some point in your life online.

Start with the assumption that even the most secure technologies are vulnerable. There's an ongoing war between criminal hackers and security experts, and that's not going to change. The only way we can "win" is to assume everything will be hacked, and take precautions to secure what is important. If you expect this inevitable hacking of your security systems, you will be able to understand the risk factors and monitor your security on an ongoing basis. You'll know the places you are vulnerable and be able to take appropriate precautions.

How to know which are the appropriate precautions? That's easy. Read this book! Many of the vulnerabilities enumerated in this book can be dealt with relatively easily, once you have the know-how. You don't need to have the most secure system, just the best one for your needs. Not sure what those are? We'll help you figure that out.

In a sense, hackers, in their own way. Every time they break a system, we learn something new about its vulnerability, and how to make it more secure. I personally look forward to the new and exciting ways hackers will point out the limitations of each new technology. I just don't want them learning on you!

HEATHER VESCENT

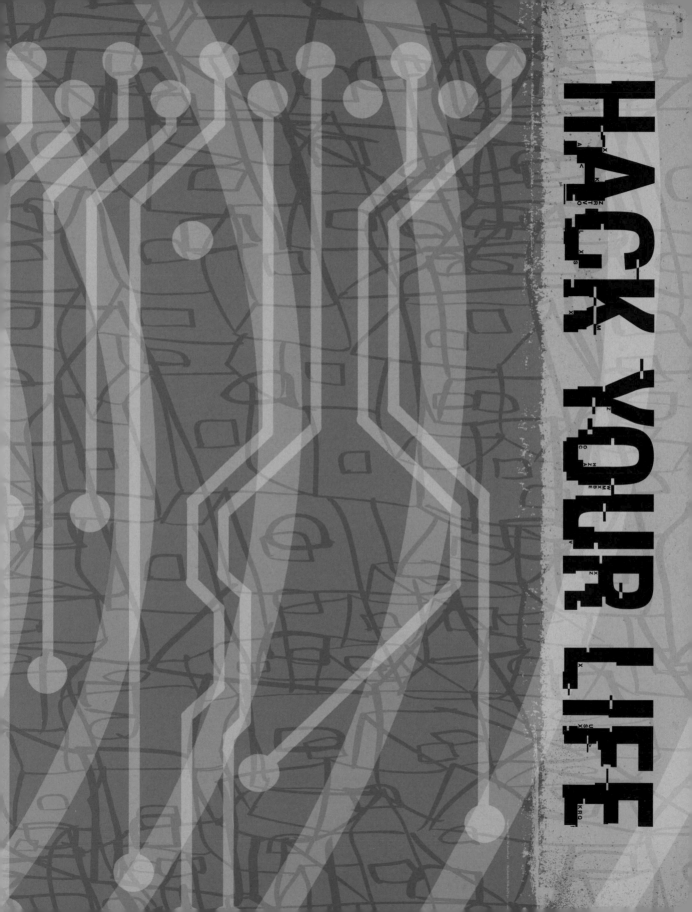

Your bank account is suddenly, mysteriously overdrawn. Everyone in your address book gets a desperate email from you asking for money. You fail what should have been a routine background check. Your TV starts getting unusual error messages. What's going on? Cybercrime can, quite literally, hit you where you live—and it's getting more common all the time as our lives get more connected and hackers more sophisticated. The chapters that follow tell you what to do when Internet bad guys make it personal—stealing your identity or your money, invading your privacy, bullying your kids, or even threatening your life. We also highlight some unexpected vulnerabilities in your smart phone, your browsing habits, and your household appliances, as well how to keep your personal information safe and secure.

KEEP YOUR IDENTITY SAFE

IDENTITY THIEVES CAN BUY, SELL, OR CAPTURE YOUR IDENTITY AND USE THE INFORMATION TO GET MONEY AND SERVICES`—OR USE YOUR NAME, CREDIT RATING, OR INSURANCE TO TAKE OUT A LOAN OR GET FREE MEDICAL CARE.

There are myriad ways for the bad guys to get your information and use it for all sorts of nefarious purposes—mainly, stealing your money, although occasionally for other kinds of fraud or to cover their tracks when committing additional crimes. That's one of the big reasons identity theft can be so devastating. If a criminal steals your credit card information, your bank will likely refund you the money that was lost. If the same criminal impersonates you to run an international child pornography ring, however, then your problems just got a whole lot worse . . . especially since many law enforcement folks aren't up on the latest types of cybercrime, so "that wasn't me" might not go over well.

How does it happen? We'll examine the many methods of identity theft in the pages that follow, and we'll also show you how you can protect yourself from being a victim or fight back if you already are. The methods of ID theft range from the seriously low tech (such as digging through your trash for unshredded financial documents or stealing those new credit cards that the bank sends you unexpectedly) to sophisticated database breaches and other hacks staged half the world away by large crime syndicates to fund cyberterrorism operations.

AMERICA'S FIRST IDENTITY FRAUD

Philip Hendrik Nering Bögel had some financial problems, and he was a creative thinker. So in 1793, when things got too hot for the Dutchman (who was wanted for embezzlement at the time), he did what any forward-thinking identity thief would do today: He hot-footed it out of the Netherlands, setting forth on this continent a new city, conceived in parsimony, and dedicated to the proposition that Bögel deserved better. Calling himself "Felipe Enrique Neri, Baron de Bastrop," Bögel started being awfully helpful to early Texas leaders Moses and Stephen F. Austin in obtaining land grants. After being named Texas land commissioner, Bögel came to settle a Texas city that he named after himself. Today, visitors to Bastrop, Texas, population 5,340, can celebrate how America's earliest successful ID fraud operation netted one guy a whole city.

T/F

"MY IDENTITY ISN'T WORTH STEALING!"

FALSE Attackers are smart, and they seek the easiest path to their ultimate target. Often, that easiest path runs through your computer is *you*. You may say, "I just have photos of my grandkids on my hard drive." But your machine is connected to the internet, making it a target. Hackers can hijack your computer and join it into a secret global network for spam, attacks on other computers, and more nefarious activities. While they're at it, they might just steal your banking information as well. It is also not unknown for hackers to destroy a computer, so that even those family photos that are priceless to you, while worthless to others, end up lost with the dead computer.

– – – – – – – – – –

MANY TYPES OF IDENTITY THEFT Criminals impersonate you online for a range of different reasons and in a variety of ways. For cyberstalkers (see pages 50-51 for Amanda Nickerson's story), the impersonation is usually part of a larger cyberbullying effort. But in most cases, the motivations are financial. Whether it's designed to get bank cards or bank loans in your name, obtain credit in your name, or impersonate you to use your existing credit, identity theft is usually a gateway cybercrime—an initial act, atop which lie other criminal schemes. So really, "identity theft" should be thought of as a family, or a category, of cybercrime.

Even though it's common for victims to be reimbursed by banks or credit card companies, the damage done by ID theft can affect you for years. Your credit score and history are the main ways that banks, car dealers, and other lenders determine the risk of extending you credit, and the black marks can be hard to erase.

A Taxing Scheme One of the fastest growing crimes in America is tax return fraud, which can net identity thieves thousands of dollars for each successful impersonation they make to the IRS. The criminals get hold of your Social Security number and personal information, and then create a tax return in your name that shows a modest overpayment on your part. The return is filed online using software, and within days, the IRS sends out a refund to "you"—at the address given by the thief. The refund is typically made using prepaid Visa cards, which can be easily exchanged for cash or property.

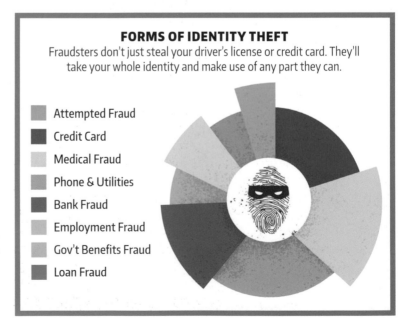

FORMS OF IDENTITY THEFT

Fraudsters don't just steal your driver's license or credit card. They'll take your whole identity and make use of any part they can.

- Attempted Fraud
- Credit Card
- Medical Fraud
- Phone & Utilities
- Bank Fraud
- Employment Fraud
- Gov't Benefits Fraud
- Loan Fraud

STRANGERS WITH CANDY In 2004, some InfoSec folks did a little experiment in which they offered passersby on the street a candy bar if they would tell them their work logins and passwords. To their surprise, some 70 percent were willing to part with the information—half of them did so even without the chocolatey bribe. You'd think that would have been a wake-up call. And indeed, governmental agencies and private-sector companies spend millions of dollars on training to make employees aware of proper security procedures and how important it is to follow them. How's that going? Well, when the experiment was repeated in London in 2008, there was no difference.

Whether the reasons are cultural or technical, the fact is, people are just really bad at keeping their passwords secret. They just don't take it seriously. What's even more galling to those who work with companies and individuals to improve security comprehension is that "your password" is still taken literally. By which I mean that most people to this day use just one password for many or all of their accounts—and a weak one, at that (see page 28 for more on creating a secure password).

You might think that this problem would have already been solved with the creation of password manager apps, which significantly reduce the toil and trouble of thinking up (let alone remembering) strong new passwords, such as the ever-popular 98cLKd2rh29#@36kasgJ!. Plus, the programs are easy to use and can automatically change the passwords for all your online accounts.

So in 2016, when a security consultant decided to try the chocolate bar trick again, this time staging it as a contest in which the person with the "best" login and password would win prizes ranging from candy to a bottle of Champagne, he finally got different results: They were even worse than before.

GUARD THOSE DIGITS You should think thrice before handing over your Social Security number (or, outside the U.S., your national identity number), even if a legitimate office is requesting it from you. This number is a universal identifier, and you've probably been asked for it multiple times a year, every time you open a bank account, take out a loan, or verify your personal information. It always pays to think about why it would be needed and to refuse to provide it unless it is absolutely necessary. If you're paying cash, never give out the number. I would rather put down a $75 deposit to get electricity or phone service than provide the utility company with my Social Security number—plenty of utilities have been routinely hacked, and ID theft in America thrives on this ubiquitous identifier. If the service provider doesn't need it, don't provide it to them.

TINFOIL HATS It's a common joke that some people are so paranoid, they line their hats in tinfoil. Funny thing? That might not always be such a bad idea.

There are many ways to conduct data theft, and some of them do rely on secret transmissions. The best (or, at least, one of the coolest) examples of this was the Soviet hack against IBM Selectric II and III typewriters in the 1970s. About fifteen of these were used in the U.S. Embassy in Moscow and the consulate in Leningrad, and were modified by Soviet spies to contain a device that measured the magnetic disturbances generated when the little Selectric ball swiveled. Each letter, it turned out, had its own signature. By implanting a receiver in the walls (the buildings were, of course, built by Soviet contractors), the government could see the very pages of documents as they were typed up.

HOW THEY DO IT Criminals engage in obtaining identities to exploit in a range of ways, from low-tech to Secret Squirrel. Once the most common method of identity theft, paper or wallet theft is still popular, but now it's a small-time operation. Still, someone lifting your wallet and using your ID and credit cards can do a fair bit of damage. Similarly, ID theft can occur when people rifle through your trash and find bank statements and other bills with account numbers, balances, and dates. These specifics allow thieves to call those vendors and report your cards as lost, change your address, and have replacements mailed to them.

Other schemes to separate you from your identity run the gamut from physical theft of personal documents from service providers to breaking into a computer network specifically for the purpose of stealing data. Another popular method is phishing (see page 24).

But of course, the most common method of stealing identities is to do so en masse in a large-scale breach of a retailer, bank, insurance provider, or government agency. This gives criminals the biggest bang for their buck and the largest number of targets. See the chart on the facing page for more information about how this works.

One Step Ahead of the Law It is very difficult for authorities to prevent or successfully prosecute identity thieves. Because much of the fraud can be done at a distance and by using online tools, catching the criminals in the act is difficult. What's more, with the global nature of the internet, the criminals don't even have to be in the United States to commit these crimes. And, finally, ID theft can go on for some time before a victim is even aware that it has happened.

HOW MIGHT YOU BE VULNERABLE? The vast multibillion-dollar cybercrime industry can be divided into three basic categories, each with its own objectives, although at the end of the day, the result is the same: You've been had. Understanding the differences, and what happens at each stage of the game, can help you stay safe. Here's how these crimes roll out.

IF YOU ARE

THE TARGET

THAT MEANS

an adversary has targeted you on a highly personalized basis.

IN THIS CASE, THE HACKER MIGHT WANT TO

extort money from your small online business.

SO HE OR SHE

crafts an email to you personally, using specific details to convince you he works for your website's registrar.

AND THEN

believing you're speaking to your own provider, you reveal the log-in information for your account.

ONCE THAT'S DONE

the hacker logs in, takes your site down, and changes your password.

AND IN THE END

the hacker demands a $5,000 USD wire transfer to restore your site.

IN THE TARGET POOL

THAT MEANS

you are part of a group being targeted by a broad-based or general attack.

IN THIS CASE, THE HACKER MIGHT WANT TO

access PayPal accounts.

buys or builds a spamming list of ten million email addresses, one of which is yours.

AND THEN

the hacker sends a fake but realistic and compelling phishing email that tricks you and other users into revealing PayPal account log-in information.

ONCE THAT'S DONE

the hacker harvests logins from anyone who fell for the phishing email.

AND IN THE END

the hacker logs into your account and sends himself a fraudulent payment.

THE VICTIM BUT NOT THE TARGET

THAT MEANS

you are a bystander caught up in someone else's mistake.

IN THIS CASE, THE HACKER MIGHT WANT TO

access health records at a major insurer.

SO HE OR SHE

registers a look-alike domain resembling the real one, say One-Health.com, instead of OneHealth.com.

AND THEN

the hacker crafts believable emails using a company executive's name, role, and title to convince users to open a malicious attachment.

ONCE THAT'S DONE

the hacker accesses the network, in this case gaining access to millions of private medical records.

AND IN THE END

your records are stolen even though you're not the one who clicked on the malware.

KEY CONCEPT

WHY IS IT CALLED PHISHING? Phishing is a term used to describe some of the most widespread and effective methods for obtaining information online. The term itself is a mash-up of two words—"fishing" and "phreak." The fishing part is just what you'd imagine: to fish for victims or data by using electronic bait, hooking victims, and reeling them in—an obvious and accurate metaphor for the act itself. The alternate spelling is a nod to the pre-internet practice of telephone-system hacking known as phone "phreaking," done by "phreaks." This is related to another hacker practice, called "1eet speak," which substitutes numbers for letters and some letters for others to create an often goofy insider jargon. It's quaint today, but you will still see versions in chat rooms, as hackers somewhat jokingly refer to one another as "133t H4x0r5," or "elite hackers."

TEACH A MAN TO PHISH Phishing isn't one specific thing. Rather, the term is used for a wide range of methods designed to gain access to your information. Understanding what those methods are, along with the basics of how they work, is central to both recognizing and avoiding many of the risks you face online. So before we go any further, let's do a quick overview of the many types of phish in the sea and the ways they can bite. Here are three common methods that these criminals will try when going after your data.

Voluntary Disclosure The first method is diabolically simple: Attackers use a rich mix of psychological techniques, known collectively as *social engineering*, to get you to give up the goods, essentially conning you into giving away the information that they want. People are generally trusting, and it's amazing how much information the average person will give up simply because someone happened to ask them in the right way.

Malicious Attachments In these cases, computer users are tricked by some compelling message into opening a poisoned email attachment, which then installs malicious malware on their machine, thus giving the hacker access to their computer or network. These masquerade as documents that the users "requested," photos they "just have to see to believe," and the like.

Malicious Links Because many email systems can now block out malicious email attachments, some attacks will use malicious links to drive the user to an infectious web page instead. Most people are so accustomed to clicking on links almost automatically that this technique is highly effective. Most of these links are disguised to boot—an image in the email with a logo or a line of text displaying an address or site to visit that is actually a cover for a malicious web address which a hacker has set up for just this purpose.

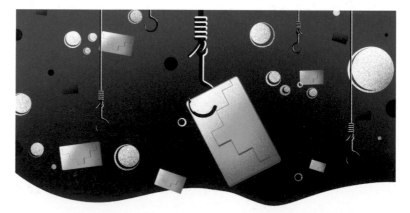

TYPES OF PHISH There are a lot of phishing schemes in the sea. You've probably been exposed to at least a couple of the examples listed below—and hopefully you didn't fall for them, although if you did, you're one of millions of people who have. Using the information below, you'll be better able to spot these scams and steer clear.

TYPE OF SCAM	HOW IT WORKS
CLASSIC PHISHING	A fake website "spoofs" or closely resembles a real one, into which users enter their access credentials, identity data, or other sensitive information.
SPEARPHISHING	As the name would imply, this is a highly targeted attack, often designed to victimize a small, specific group or even one individual, using highly personalized messages that may be the result of hours or even weeks of online reconnaissance on the target.
WHALE PHISHING	The spearphishing of a high-profile or high-value individual, such as a CEO or celebrity, that is, a "big fish" or whale.
CATPHISHING	The use of fake online personas or profiles to create a phony emotional or romantic relationship, either for financial gain or access to sensitive information.
VISHING/SMISHING	Scams or data thefts that leverage phishing-like techniques but target phone users over voice lines or SMS.

IF YOU'RE ENTERING PERSONAL DATA ONLINE, TYPE THE ADDRESS YOURSELF AND CONFIRM THE SITE IS SECURED WITH AN HTTPS PREFIX AND A CLOSED-LOCK ICON.

PHISHING EMAILS ARE EASY TO DETECT

FALSE A lot of people believe that they can easily tell when they're being phished through email. But more and more often, scammers are crafting messages that appear to be from a legitimate source, such as your bank or your Amazon or eBay account, complete with a full page of images and icons from those sites duplicating a genuine email—but secretly redirecting an unsuspecting user to another site. You can sometimes confirm it's a fake by moving your mouse over the link (without clicking) and seeing another address pop up in preview. But just to be on the safe side, you should always enter the address yourself, never by clicking links.

- - - : - - -

DEFEND YOURSELF AGAINST PHISHING So if the thieves are smart, and not even the rich and famous can protect themselves, does that mean you're hosed? Not at all. That's because in most cases, victims fall for these attacks not out of a lack of resources but a lack of awareness. An astute and informed user with a zero-dollar budget is harder to victimize than an oblivious and untrained one with all the money in the world. Here are five simple steps you can take, starting right now, that will make you a significantly tougher target for phishers.

Be Aware Simple awareness is the first line of defense. Be suspicious. Understand and believe that you are a target. Treat any message in any electronic medium from someone you don't know as highly suspect.

Use the Hover Test Any modern email program will show you the destination of a hyperlink if you mouse over it without clicking. This "hover test" can help you spot suspicious links in any email you've received. If the visible link and the underlying destination don't match exactly, don't click!

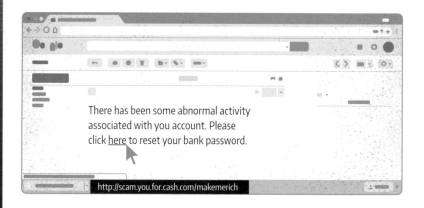

There has been some abnormal activity associated with you account. Please click here to reset your bank password.

http://scam.you.for.cash.com/makemerich

Check the URL Learn how to properly read a web address. The name of the site you're visiting is the last thing to the left of the first single slash, not the first thing to the right of the double slash. Phishers constantly use this lack of knowledge to trick people.

> **SAFE: https://www.amazon.com/**
> **UNSAFE: http://www.amazon.phishingforyou.com/**

Be Attachment Phobic Malicious attachments are the number one way to let password stealers, Trojan horse viruses, and other nasties get onto your computer. You should only open attachments from people you know, and even then limit yourself to messages you're expecting, such as an invoice for services you actually have received.

Confirm Out-of-Band If you happen to receive a suspicious message or a request for information that seems too personal, even from individuals or companies you trust, confirm the request via a different medium. For example, if they email you asking for your information or requesting that you click the link to their website to correct an issue, try visiting their website or calling them by phone. And remember, type the web address out manually or find the phone number yourself. Never rely on the link or phone number in the suspicious message. Those could both be fakes run by the phisher!

HACKER HISTORY

PHONING IT IN The first known online mention of the term "phishing" was in the online group alt.2600, a discussion forum for phone hackers, in early 1996. The "2600" refers to the frequency in hertz that early phone phreakers discovered they could play into a phone handset to take over the phone company's switches and make free calls to anywhere in the world. That this hack was so simple to execute, and so fundamental to the system that it was simply too expensive to fix, led to an entire subculture around building "blue boxes," or tone generators that would play the 2600 Hz whistle tone. Even Steve Jobs and Steve Wozniak, of Apple fame, sold them in the early days. One intrepid phreaker, John Draper, worked with some blind phreakers who were, as you'd imagine, particularly sensitive to tone. He learned that a plastic whistle offered as a free prize in boxes of Cap'n Crunch cereal blew at, yes, 2600 Hz. Draper used the whistle widely and became known in hacking circles as Crunchman. He's still around, too: You can find him on Twitter @jdcrunchman, or look for John "Captain Crunch" Draper on Facebook.

GOOD TO KNOW

YOU'RE NOT ALONE Millions of ordinary citizens have been victimized by one type of hack or another. Even the smart, powerful, and rich have been victims. For example, real-life rocket scientists at NASA have had their computers taken over by Chinese hackers. The U.S. government has concluded that Russians hacked the DNC and that Anonymous hacked Donald Trump during the 2016 election. In 2008, vice presidential candidate Sarah Palin's email was stolen by a hacker who figured out the Alaska governor's email password. Other notable victims have included Attorney General Eric Holder, FBI Director Robert Mueller, Jay Z and Beyoncé, Paris Hilton, Mel Gibson, Kim Kardashian—and Nick Selby, one of the authors of this book. This isn't even taking into account the massive amounts of top-secret government information released by WikiLeaks, Edward Snowden, and others.

CAN I TALK TO YOUR MANAGER? The longest, most complex passwords are impossible for hackers to break in a lifetime (or even several!), but it also seems as if they might take a lifetime to come up with and nearly as long to input each time you have to use them. Luckily for you, there are password manager programs out there that can do all of the heavy lifting for you.

A password manager site or application like LastPass, Dashlane, or 1Password can generate, store, and encrypt a list of passwords for you, import any passwords that you have previously created yourself from browsers, analyze the strength of a password, and more.

Just be sure that you can remember and keep secure the master password to the account itself—and luckily, many password managers also offer two-factor authentication (see facing page) for an added layer of password protection.

CREATE A POWERFUL PASSWORD Now that you know what to avoid in emails, what's the next step? Well, every online account requires an account name (often derived from your own name or email address) and a password. The following guidelines can help you come up with passwords that are as unbreakable as possible.

One Size Does Not Fit All Look at the keys on a key ring: Each is a different design and cut. Just as each key is made to fit a specific lock, each password should be unique to the account it's used for. Otherwise, if you're a victim of ID theft, whoever stole your information will have access to every single account of yours that the criminal can think to try.

Bigger Is Better Some sites limit how long your password can be. While a long password may be hard to remember, it's harder for a hacker to break, even with brute-force methods (that is, using programs that try every single possible combination of characters).

Get Complicated Passphrases like "correcthorsebatterystaple" are easy to remember, but anything that uses dictionary words is easily hackable. Avoid simple substitutions, too, such as "p4ssw0rd" instead of "password." Use every single type of character you can: lowercase and capital letters, numbers, punctuation, and anything else available. Finding a number between 0 and 9 is easy for a hacker or ID thief; finding the right character in a total of sixty-two numbers and lowercase and capital letters is massively more challenging, especially the longer the string gets. If you have to write down a password to help remember it, keep said document hidden and safe from prying eyes or theft, or consider using a password manager.

Change Is Good Don't just come up with a password and then leave it be. Change your passwords frequently and, if at all possible, never reuse one. If hackers steal older data, they may score a hit if you're using that old password for a new account.

JUST DON'T The top ten most common—and thus worst—passwords have stayed largely the same since passwords became a thing, only changing in order from year to year. Right now the top contenders are:

1.	123456	6.	password
2.	123456789	7.	123123
3.	111111	8.	000000
4.	qwerty	9.	1234567
5.	12345678	10.	1234567890

WHO WANTS TO KNOW? Sometimes an extra layer of protection, called "knowledge-based authentication," or KBA, is added to your password, either in addition to your basic login and password or to verify your identity if you've forgotten your password. Of course, like many other defenses, this tool can also be turned against you.

Static KBA Also known as "shared secret questions," these are questions along the lines of your mother's maiden name, town where you were born, and so forth—often matters of public record. In addition, this information is stored somewhere, so it can be stolen, which means that even the weirder questions, like "Who's your favorite poet?" aren't secure.

Dynamic KBA Here, questions are generated in real time from a range of public and private records. You don't know what questions will be asked, but, hopefully, you'll remember the answer. Examples might include "What color was your Honda Accord?" or "Which of these streets have you never lived on?" You only have a short time to answer; the odds of someone guessing correctly on the fly are lower.

Unfortunately, you may not have the luxury of only patronizing sites with excellent dynamic KBA, although if you have the choice, take it. The simple workaround? Lie. It's relatively easy to figure out where someone went to high school. But if the "correct" answer is Narnia, Petticoat Junction, or Westeros, that's less likely to show up in old yearbooks. More's the pity.

USING THE POWER OF TWO Two-factor authentication, also called "2FA," is like a counterpassword in a spy novel ("The blackbird sings at midnight"; "But only under a full moon") or two people turning keys at the same time to launch a missile. Often available as a mobile app or a physical token (something like a key ring tag) that only you would have access to, 2FA uses a shared algorithm attached to your account. After typing in your password, you're prompted to use the app or push a button on the tag to generate the authentication key based on that algorithm, usually a short string of numbers randomly created on the spot. If an account offers 2FA (such as Google Authenticator), use it, and your accounts will be that much harder to compromise. If you should lose the token or the mobile device with the app, replace it ASAP so you can keep your account safe.

IDENTITY THEFT

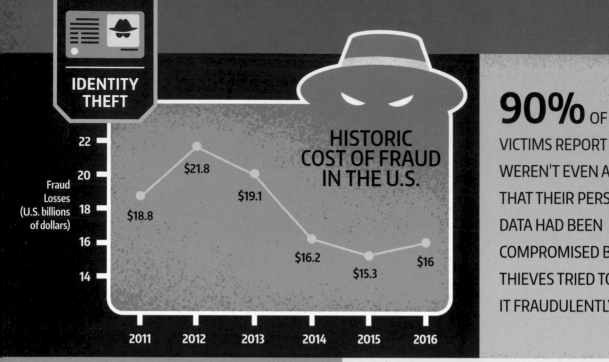

HISTORIC COST OF FRAUD IN THE U.S.

Fraud Losses (U.S. billions of dollars)

- $18.8 (2011)
- $21.8 (2012)
- $19.1 (2013)
- $16.2 (2014)
- $15.3 (2015)
- $16 (2016)

90% OF VICTIMS REPORT THEY WEREN'T EVEN AWARE THAT THEIR PERSONAL DATA HAD BEEN COMPROMISED BEFORE THIEVES TRIED TO USE IT FRAUDULENTLY.

GROWTH IN CASES OF ID FRAUD

Millions of U.S. Fraud Victims

| 11.6 | 12.6 | 13.1 | 12.7 | 13.1 | 15.4 |

THE MOST TARGETED GROUPS

PEOPLE AGED 18 TO 24

ANYONE WITH A WEAK PASSWORD

1234

YOUNG CHILDREN

PEOPLE MAKING OVER $75,000

COUNTRIES WITH HIGHEST INCIDENCE OF IDENTITY THEFT PER CAPITA

32% of victims do not notify the police

1. MEXICO
2. THE USA
3. INDIA
4. THE UNITED ARAB EMIRATES
5. CHINA
6. THE UNITED KINGDOM
7. BRAZIL
8. AUSTRALIA
9. SINGAPORE
10. SOUTH AFRICA
11. CANADA

64% Credit cards that showed attempted or successful use by thieves.

IF YOUR PERSONAL INFORMATION IS STOLEN, HOW LONG DOES IT TAKE TO RESOLVE?

48%	1 DAY OR LESS
16%	2 TO 7 DAYS
18%	8 DAYS TO A MONTH
6%	1 TO 3 MONTHS
3%	3 TO 6 MONTHS
9%	6 MONTHS OR MORE

SIGNS YOUR IDENTITY HAS BEEN STOLEN

YOUR CREDIT CARDS RATES GO UP

YOUR CREDIT RATING DROPS

MYSTERY BILLS SHOW UP

CAR INSURANCE GOES UP

BANK STATEMENTS STOP COMING

TAX REFUND DENIED

CAN'T RENEW DRIVERS LICENSE

CHECK-UP REMINDERS FOR MEDICAL CONDITIONS YOU DON'T HAVE

CALLS FROM COLLECTION AGENCIES

TURNED DOWN FOR LOAN

FAIL TO PASS BACKGROUND CHECK FOR JOB

30 DAYS
Average time needed to handle identity theft crime.

MOST COMMON FORMS OF IDENTITY THEFT

49.2%	15.8%	9.9%	5.9%	3.5%	3.3%	22.9%
ATTEMPT TO ACCESS GOVERNMENT DOCUMENTS*	CREDIT CARD FRAUD	PHONE OF UTILITIES FRAUD	NON-CREDIT CARD BANK FRAUD	LOAN FRAUD	EMPLOYMENT-RELATED FRAUD	MISC OTHER FORMS

*ID, social security number, tax returns, etc.

30 HOURS
Average time to handle and settle a disputed charge with a credit company.

SECURITY BASICS

Your wallet often has all of your identification and bank cards (and more). If that wallet gets stolen, your entire life's identity and finances will literally be in someone else's hands. Should that happen, the best plan is to have culled its contents well beforehand so that you're only carrying the minimum number of IDs and credit cards—nothing more than is absolutely necessary. This will limit your losses in case of theft. And, it means that the only calls you will have to make will be to your credit card company, your local DMV office, and your employer to report the losses. Your credit card and driver's license will be replaced, and your employer can deactivate your work ID card, thus preventing whoever stole your wallet from using the card to break into your office and clean you out of paper clips and printer ink cartridges.

THE SEVEN-POINT ID THEFT RECOVERY PLAN If you have been the victim of identity theft, it is very important that you take steps to safeguard your good credit, warn the appropriate agencies of the event, and protect your good name. Often, you'll want to talk to the police. That's a good idea, but don't be surprised if you learn that there's not a lot they can do. The rest of this chapter explains how you can help yourself when you are the victim of identity theft. If you don't, it can cost you dearly when applying for a car loan, mortgage, or credit card. It could also make it harder for you to find a job, rent an apartment, or buy insurance.

The first thing you must do when you are a victim of identity theft is to get organized. The seven-step checklist here is just a suggested series of steps; customize it as necessary to your needs.

STEP 1	FILE A POLICE REPORT
	If you discover you have been victimized, contact the non-emergency number of your local police department and ask to speak to a detective.

STEP 2	GATHER DOCUMENTS AND EVIDENCE
	Contact your nation's consumer protection agencies, as well as stores and creditors to gain copies of the documents used to open accounts in your name.

STEP 3	CREATE AN AFFIDAVIT AND ID THEFT REPORT
	Your local consumer protection agency should be able to provide documents you will need and demonstrate how to present them. They also provide sample forms for an identity theft report, which, along with your police report, will help speed up the process with creditors, banks, and other agencies.

STEP 4

INFORM THE CREDIT AGENCIES AND CREATE AN EXTENDED ALERT

To establish a fraud alert with the credit agencies, contact them directly. You will need to reissue the alert every ninety days.

STEP 5

INFORM YOUR BANK, CREDITORS, AND MERCHANTS

With the package you've created, contact your bank and other creditors and merchants with whom you have accounts and inform them of the issues you have faced.

STEP 6

PROTECT YOUR SOCIAL SECURITY NUMBER

If your number was misused, inform the national agency and request information on an ID Theft Affidavit. You may also wish to contact your agency if your Social Security number is being continually abused or phone to victims of identity theft.

STEP 7

MONITOR YOUR CREDIT

You are entitled to at least one free credit report per year, but that is often insufficient for monitoring. There are several commercial companies offering these services, and we recommend you seek professional advice on which to choose. Several nonprofit organizations are out there to help victims, offering assistance to victims of identity theft by internet or phone.

SHRED AND THOROUGHLY DESTROY ANY OLD AND UNUSED CREDIT CARD APPLICATIONS OR SIMILAR FORMS TO HELP KEEP YOUR INFORMATION OUT OF THE HANDS OF ID THIEVES.

SYNTHETIC ID THEFT This chapter deals with the theft of someone's actual identity, but here's a new twist: synthetic identity theft. That's when an identity that has never before existed is created by scammers. Identity thieves typically seek to obtain names, national identity numbers and dates of birth, medical account numbers, addresses, birth certificates, death certificates, passport numbers, bank account or credit card numbers, passwords (like your mother's maiden name or children's or pet's names), telephone numbers, and even biometric data (such as fingerprints or iris scans). With synthetic ID theft, thieves only need some of this information to create a whole new fake person.

Thieves then create a credit file—the closest thing in the digital domain to conjuring up a human. This exploits a weakness in the authentication scheme used by credit reporting agencies: If an identity doesn't exist when it is checked, a new file is created. And a file? That's gold.

Credit Where No Credit Is Due The best thing to do with a synthetic ID is build its credit over time. This can be done in the traditional way—almost anyone can get a high-interest, low-limit, unsecured credit card at a hardware store, so the idea is to get one, then buy a hammer and pay it off over time. To get fancier about it, they might join up with a "data furnisher" who works at a business and will write up a phantom credit account for our spooky friend, showing scheduled payments made over time to speed things up. There's an entire industry around this, because the stakes are very high.

The most common way is to conjure up children. This is because, for the eighteen years or so after most kids are born, they don't do anything with their credit. During that time, anyone who establishes a credit file for the young one in question would likely be free from

> IF YOU HAVE CHILDREN, PRIVATIZE AS MUCH OF THEIR INFORMATION AS YOU CAN, AS THEY ARE THE GROUP MOST VULNERABLE TO SYNTHETIC ID THEFT.

any interference until someone notices—that's typically at just about the worst time: when the kid applies for a college loan. The best way to protect against misuse of your child's credit is the same as it is for yours: Check it regularly, and check on it as often as you can. Should you happen to see fraudulent accounts, yell early, often, and loudly.

If you are on active duty in the military, it is recommended that you put an active duty alert on your own credit files by contacting any one of the three major credit agencies. Credit agencies all share active duty alerts. Each alert will stay in your files for at least twelve months. If someone applies for credit in your name, creditors will take extra precautions to make sure that the applicant is really you.

THE TAKEAWAY

Here's how to apply the lessons of this chapter, whether you're looking for basic safeguards, enhanced security, or super-spy measures to safeguard your privacy.

BASIC SECURITY

- Use strong passwords.
- Use different passwords for every site.
- Use a password vault program.
- Never share your login information with anyone.
- Don't click on suspicious links or download unexpected files.

ADVANCED MEASURES

- Always use two-factor authentication.
- Don't get kids social security cards unless necessary.
- Check your kids' credit at least quarterly.

TINFOIL-HAT BRIGADE

- If any service provider's site uses weak KBA, take your business elsewhere.
- File your taxes the old-fashioned way, on paper.
- Eschew electronic information wherever possible.

GOOD TO KNOW

WHAT LAWS PROTECT YOU? In virtually every place you care to look, identity theft is considered a federal crime. But it can still be next to impossible to actually get a federal office to investigate your individual case of identity theft—well, unless you are famous, or rich, or there is something larger at stake connected to the theft itself. Most states have their own laws against identity theft as well, and your local police department may have a program that can help you—ask them what resources are available in your area. Ultimately, however, you may simply be on your own, as it can be difficult to track down a specific perpetrator of identity theft (especially given that you may just be one of many victims caught in the same sweep). Usually, the best you can do at the local level is work to limit the damage done and clear your name.

WHERE THE MONEY IS

THERE IS MONEY IN CYBERCRIME. NOT "BUY A NEW CAR" MONEY. NOT "BUY A NEW HOUSE" MONEY. THIS IS "BUY A NEW AIRPLANE" MONEY, MAYBE EVEN "BUY AN ISLAND" MONEY. OF COURSE, IT'S MONEY THAT BELONGS TO YOU.

In his autobiography *Where the Money Was: The Memoirs of a Bank Robber*, America's most celebrated bank robber, Willie Sutton, denied ever saying that he robbed banks because "That's where the money is." Nonetheless, it's a great line. And for cybercriminals, it's a directive. Where is the money these days? On the internet. Even garden-variety spammers and botnet managers can expect to bring in $20,000, $30,000, even $50,000 USD a week. If you're bad at it, you'll make less. If you're good at it . . . well, the FBI said that the hacker and criminal-empire builder known as Dread Pirate Roberts was earning $1 million USD per week before his arrest. That's seven dollars and fourteen cents every second.

And you don't even have to be a criminal to pull down big bucks from hacking—even the so-called "white hat hackers" (also called "ethical" hackers) can have a payday, too. The FBI is said to have paid a cool million for the hack that enabled the bureau to access the iPhone belonging to one of the suspects in the 2015 San Bernardino mass shooting, and in September of 2016, *Wired* magazine reported that a high quality, previously unknown iPhone hack had been sold for $1.5 million USD.

In short, both bad guys and good guys hack . . . because that's where the money is.

SUCH A NICE BOY By all accounts, Maksym "Maksik" Yastremskiy is a nice boy. The twenty-five-year-old baby-faced Ukrainian was fun and friendly, and he bought nice things for his mother. He could afford to because in just about a year, young Maksik cleared some $11 million USD selling credit cards he and his friends stole from T.J.Maxx, an American retailer, before he was busted. Maksik's accomplice, Albert Gonzalez, was arrested as well. Perhaps suspicious of banks, Gonzalez had buried a barrel containing $1.1 million in cash in the backyard of his parents' home. Cops seized other fruits of Gonzalez's work, in the form of a new BMW, a condominium in Miami, his ex-girlfriend's Tiffany diamond, and three Rolex watches. So great were the stacks of cash he netted, Gonzalez bought himself the kind of currency counter used by bank tellers and casino cashiers.

THE ELECTRONIC ECONOMY When the average person thinks of the economic side of cybercrime, what comes to mind is theft . . . someone stealing your credit cards or other funds electronically. And, indeed, this is a massive business, with some $15 billion USD stolen electronically every year. However, there are other sketchy ways that money changes hands (or flies out of your wallet) online. In this chapter, we'll examine a number of them—and how to protect yourself.

Identity Politics Often times, a phishing expedition or other sort of identity theft is just the first step in a series of attacks. While an identity thief may use data stolen from you for a number of purposes, as discussed in the previous chapter, the most common is to steal your money or use your identity as a shield for a larger theft. That's why it's so crucial to do your due diligence when you discover identity theft or fraud. After all, your credit card will almost certainly refund any fraudulent charges as long as you promptly report the card missing and file a police report, if required. However, if the thief then goes on and uses your identity to front a multimillion-dollar international con game, that would be a little less easy to resolve with a call to your local customer service rep.

Shady Sales Criminals don't have to steal your identity to get their hands on your money. You might be willing to hand it to them with a smile. We'll talk about the deep end of unofficial online markets in future chapters, but know that you don't have to be buying an AK-47 or a kilo of heroin to be part of the underground economy. It can be much more mundane on the so-called gray market.

NEVER REPLY TO ANY EMAIL THAT CLAIMS TO HAVE BEEN SENT BY YOUR BANK AND IS ASKING FOR YOUR ACCOUNT INFO. INVARIABLY, IT'S A SCAM.

DATA THIEVERY

There are a multitude of ways that criminals can steal your data, from hacking into computers to pulling confidence scams.

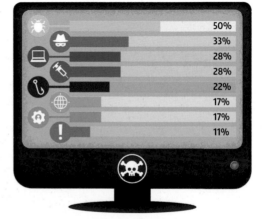

- Viruses, malware, worms, trojans — 50%
- Criminal insider — 33%
- Theft of data-bearing devices — 28%
- SQL injection — 28%
- Phishing — 22%
- Web-based attacks — 17%
- Social engineering — 17%
- 11%

NOT THAT KIND OF TRADING CARD As we were finishing writing this chapter, Nick's wife had data skimmed from her credit card. Since she checks her bank records regularly, she caught it the next day. The online history showed what are called "test swipes"—some $1 transactions, followed by a $49 test (many shops don't run authorizations or need a signature for purchases under $50), and soon after, a $599 purchase from Bed Bath & Beyond.

Her card data (the information in the magnetic stripe on the back of the credit card) had been grabbed by a fraudulent card reader. This data, known as a "dump," was gathered up with others into a pack and sold to a thief. This person downloaded the data and encoded it to another magnetic strip, such as the one on a hotel card key, which then would swipe just like the original card. Once we cancelled the card, the person using the stolen data dump breathed a sigh of regret, tossed the now-dead card on the floor, removed another from a stack of about a hundred they'd encoded from different numbers in the pack, and swiped again.

ELDER FRAUD A particularly cruel form of cyber theft targets the elderly. Every year, American senior citizens lose $2.9 billion to financial fraud. A study published by the National Health Institutes concluded that, basically, the older you get, the more susceptible you are to scams. Age was a stronger predictor than financial acumen, wealth, education, or health.

That's why so many online scams target the elderly. The over-seventy set is less likely to be computer savvy and thus falls prey to the "tech support phone call" scams, in which a helpful young man informs you that your computer has been malfunctioning; for just $29 USD, he can fix the problem. All he needs is a credit card number. This scam also manifests in a more aggressive form, as an "IRS auditor" calls to demand an immediate payment on mysterious back taxes. Wherever in the world you're located, you can find helpful tips on spotting and fighting common scams at the American Association of Retired Persons' website, updated monthly as new scams emerge.

FUN FACT

WHAT'S MY IDENTITY WORTH? In brief: not much, despite the damage fraud does to you. The aftermath of an identity theft can take around thirty hours' effort to repair, with losses averaging about $4,000 USD. But to an ID thief, you're just a drop in the bucket. A single dossier of a person's full financial and personal info might sell for about $1,300 to an interested buyer, but identity thieves often buy and sell in bulk. Files full of medical insurance info cost maybe a little more than $5 per stolen identity, while tens of thousands of Social Security numbers can be sold for as little as a penny per victim—so that list of 100,000 stolen SSNs (including yours?) goes up on the black market for just a grand.

FIFTY SHADES OF GRAY MARKET Most of us have at least heard of black-market goods. But what about the gray market? Both gray- and black-market goods are purchased outside of the usual channels, compared to white-market (totally legit) merchandise. The difference is legality: Gray-market items may be technically legal, such as single items sold from a bulk pack or merchandise sent from a cheaper to more-costly region (such as from Scandinavia to the United States).

Gray becomes black, and thus totally illegal, when goods are faked or counterfeited. It's often hard to tell whether you're getting the real deal or if that "Viagra" contains an overdose of the drug or nothing but blue printer's ink and plaster. Some black-market items are even malicious or actively harmful and thus considered red-market.

WHAT ELSE IS ON SALE? A comprehensive list of all the merchandise in each market out there would take up volumes. But here are some examples of each (we'll get into the more serious stuff in Chapter 12).

White Market: Taxed, licensed products and services; whatever you find on your local stores' shelves or on mainstream commercial sites like Amazon

Gray Market: Untaxed legitimate products; taxed but unlicensed products; smuggled cigarettes or medicines, used merchandise resold as new; certain imported vehicles

Black Market: Illegal unlicensed sales: counterfeit merchandise, stolen identities, drugs, weapons, fake IDs

Red Market: Illegal offerings actively causing physical harm: murder-for-hire, arms trafficking, slavery, child pornography

I KNOW A GUY WHO KNOWS A GUY That new game you wanted for your console this holiday season? It's now so in demand that it's on back order at the local department store, but your buddy knows a guy who works in the stockroom, and he knows when the next shipment is coming. What about that all-important textbook for your daughter's last college course? It's nearly worth its weight in gold at the university bookstore, but her roommate got a copy from Indonesia on eBay at a fraction of the cost! Maybe you need a new fridge, and you can't afford to shell out full price, but your friend says he can get you one, cheap—it just fell off the back of the truck when being unloaded is all . . .

These, in a nutshell, are gray-market goods: They're still legal, and you're still paying for them, but you're not exactly getting them through regular channels.

Just a Little Shady It's not quite a crime to possess or buy gray-market goods. Most goods for sale through these channels have been rerouted from different markets (so taxes may not have been paid on them), and trying to find them is sometimes sketchy, but they're the genuine article nonetheless.

Staying Safe Most gray-market merch is perfectly normal, but some shadier dealers will sell pre-owned items as brand new or offer defective merchandise without letting you know. If your "new" game console conks out on you, you're out some scratch and probably more than a bit frustrated—but if that scuffed-up box of "new" brake pads turns out not to give your brakes enough grip at the wrong time? *Caveat emptor* is the phrase that comes to mind.

GOOD TO KNOW

TOO GOOD TO BE TRUE The internet is full of knockoff and counterfeit goods. Here are some common examples.

Clothing and Fashion Accessories Because they are low tech and easy to make, brand-name fashions, watches, handbags, and other often-pricey accessories are massively counterfeited all over the world.

Footwear By some estimates, when you combine fashion and athletic shoes, brand-name footwear is the most counterfeited product category in the world, with fakes making up as much as one pair in ten worldwide.

Consumer Electronics A smartphone or PC might seem like a hard thing to copy, but it isn't for the thousands of firms that supply the same parts to the legitimate manufacturers.

Health and Beauty Products Gray-market sales of health and beauty aids run to 20 percent of authorized sales in most markets . . . and as high as 50 percent of authorized sales in some. That may seem harmless, but knockoff makeup and toiletries can cause severe allergic reactions, so shop accordingly.

T/F

TONER IS WORTH MORE THAN GOLD

TRUE Toner cartridges for your laser printer are ridiculously expensive, and most of what you're buying is the cheap plastic casing. The cost, and the value to the consumer, is in the few ounces of toner inside. Online fakes routinely sell for 10 to 20 percent of retail but could destroy your expensive printer. Don't risk it. Another example of these economic forces is that of cigarettes. Highly taxed, simple to produce, and high-value by weight, cigarettes are perfect for counterfeiting. Do counterfeiters take advantage of this? Well, consider that border authorities in the UK intercept an average of one million counterfeit cigarettes— every single day.

TAKING A GAMBLE So, what if you see an item that's almost certainly too good to be legit, but you're willing to turn a blind eye to the writing on the wall for a really, really good deal? From an ethics standpoint, you're on your own. We're not going to tell you it's ever okay to rip off the original manufacturers or sellers—after all, if everyone just photocopies this book and sells the copies on eBay, we'd be out of work. That said, here's the spectrum from "don't do it" to "really, really don't do it." The answer actually depends on what you're trying to avoid. Here's a good framework for how to think about making safe, informed purchases on the internet.

Rolling the Dice If you don't mind buying a cheap knockoff of designer fashions or accessories, you'll find a wealth of them online. Just understand that copies can range from totally worthless junk all the way up to identical goods made by contract suppliers on the same production line as the originals. If you understand the risks and feel like taking a chance, the worst you'll do is waste some money on a really obvious fake "Katey Spadde" handbag.

Watch for Counterfeits Some items are so prone to counterfeiting and knockoffs that, if you must buy online and want to ensure they're real, buy only from reputable sellers. Shoes from Zappos, toner from Staples, car parts from AutoZone, or CDs from Amazon are likely fine due to the strict controls used by these major retailers. The same goods from unknown sellers on eBay or Alibaba are almost certainly fake. That's okay for stockings, less so for your auto parts.

Never Ever Some things you should just never buy sight unseen online. This list includes significant assets, such as cars, boats, real estate, and so forth, which likely won't exist when you try to claim them; high-end jewelry; and prescription drugs or anything else your health or life might depend on.

GRAY-MARKET ELECTRONICS MAY SEEM LIKE A GREAT DEAL, AND THEY CAN BE, BUT THINK LONG AND HARD ABOUT HOW MUCH YOU'RE GIVING UP BY NOT GETTING THE WARRANTY, SERVICE AGREEMENTS, AND SO ON.

BIG SCAMS One astonishing blunder was made not too long ago by an English bank that decided to save money by encrypting just the "sensitive" parts of its database. So, instead of properly safeguarding it all, they only did the "account number" and "date of birth" fields and such.

And then thieves broke in through the bank's online banking application and sucked down the whole database. About three weeks later, customers began receiving letters on beautiful, cream-colored bank stationery, addressing them by name and referring to their account with that began with the numbers 271 (hint: all that bank's accounts did). "Dear customer," it read, "We value your business and want to make your online banking experience as good as it can be. Enclosed is a CD-ROM to help you! Just place the CD-ROM into your computer . . ." Of course, the CD-ROM was a combination of keylogger, fake-online-banking, and man-in-the-middle applications.

Another big-bucks scam relies on employees being scared of offending a top executive. Because bosses can't stop posting to social media, like, ever, scammers can track their movements through updates ("Just got to Shanghai, great meetings with ProX. Check out these offices!"). An employee then gets an email reading "Hey, Louise," I'm here in Shanghai and I just met with ProX Printing. Apparently, we didn't get their invoice three months ago and they are furious. Please send a wire transfer first thing this morning to . . ." The then boss provides a banking routing number and an account number. This scam is a highly effective spearphish, because it relies on so many things that seem like private data but are actually public. It works. Often. To prevent it, ensure that wire orders always—always!—follow the same verification process, by voice and with backups like second authorizers.

MOBILE BANKING As easy as it is to use your mobile phone to check your bank balance, pay bills, or transfer funds, you should still be wary, or even dispense with doing mobile banking entirely if you can. Aside from the obvious risk of losing your phone or having it stolen with any pertinent personal info on it (especially if you happen to have left it unlocked and unencrypted), there are two major issues with mobile banking: The apps offered by most banks do not support two-factor authentication (see page 29), and furthermore, many of the apps will accept any sort of security encryption info—even false info that a hacker can use for a man-in-the-middle attack on the bank's security (wherein a hacker intercepts, alters, and relays information sent between you and your bank)—and thus the security of your own account as well.

YOUR GUILTY SECRET

One gambit that surfaces every so often is the story of the "hitman with a heart." It goes something like this:

"You don't know me, but I am writing to you because, even though I am a professional hitman with scores of kills to my name, I feel sorry for you. Don't even bother trying to trace this email or going to the police; it won't work. All you need to know is, someone who knows you wants you dead. I have been paid $5,000 to kill you. But you're such a good person, so I want to give you a chance. I was given $2,500 down, and I get another $2,500 after you have been 'taken care of'. But I'll make you a deal: If you pay me the $2,500, I will simply go away. I will not kill you."

You have to wonder: do these crazy schemes actually pay off? Given that they pop up over and over, they must pay off often enough that some people keep trying them, it would seem.

LOSING CONFIDENCE Historic confidence games, such as the Spanish Prisoner, were the inspiration for a deluge of emails that flooded the early internet. In the pre-internet days, this sort of con took some effort and time, and of all those envelopes mailed out by con men only a very small percentage found a gullible mark with money to spare. The internet changed everything—it turns out the only barrier was one of scale. Suddenly, hundreds of thousands, if not millions, of emails could be sent with very little effort or cost, and the 1 percent hit rate went from the occasional celebration to a sustainable business model.

The Nigerian Prince The most enterprising of these scammers were located in Nigeria. Spurred on by the internet, an exotic-sounding locale, and some early success, boiler room operations sprung up throughout that country, with dozens of employees acting as princes. The Nigerians became so synonymous with these kinds of scams that most people in the business still refer to them as "419 scams"—419 being the chapter of the Nigerian criminal code that bans fraud.

Stranded in London One modern scam takes advantage of how common global travel has become. The "Stranded in London" gambit begins with someone hacking into your email address book and harvesting all of your contacts. Each is then sent an urgent message saying that while on a trip to London you were arrested or mugged or injured and hospitalized. The story varies but always ends with a desperate plea for the recipient to wire money immediately. The same virus that steals the contacts also shuts down the email account, so you don't see the emails from concerned friends and family asking whether you're okay, how you got to London, and which hospital you're in. Versions of this are also used after takeovers of Facebook accounts.

The Spanish Prisoner On March 20, 1898, the *New York Times* warned Americans of a new scam: It appeared that a "robber and a humbug" was sending letters to Americans from Barcelona. The writer was in prison on political charges, but, thankfully, through hard work and thrift he had managed to squirrel away $130,000. Now, he wishes to enlist the help of an honest American, you (of whom he learned through a mutual friend of great character, whose name he will, out of an abundance of caution, not mention), to help spirit this sum to America so that his beautiful daughter can marry her true love. If only you would facilitate this transaction, a third of the sum is yours to compensate you for your time and difficulty. If you could, by placing in escrow a mere trifle—say, $100—to show your good faith, the transaction can proceed.

If this sounds familiar, it's because the Spanish Prisoner is the basis for the entire family of confidence swindles known as "advance-fee fraud." As you can see, this isn't exactly new—the *Times* article pointed out that the scam was—in 1898—already an old one.

THE TAKEAWAY

Make sure your money stays in your pocket (or your bank account or online wallet) by taking the measures below—at the very least you must apply the basics.

BASIC SECURITY

- Follow up on mystery bills or collection calls immediately.
- If you lose your wallet, report aLL cards missing immediately.
- If you get a text or email from your bank asking you for info, call a branch to make sure it's legit.
- If a get-rich-quick scheme seems too good to be true, it almost certainly is.

ADVANCED MEASURES

- Check your credit report regularly.
- File a police report after fraud of any amount.
- Only use CHIP-and-signature cards (or CHIP+PIN when available).

TINFOIL-HAT BRIGADE

- Don't use banking apps on your phone
- Don't shop online.
- If a store only has swipe machines, take your business elsewhere.

CONFIDENCE SCHEME
The idea behind every one of the scams you'll find within this chapter, from the historic to the modern, from the in-person grifter to the fictional Nigerian banker or prince halfway around the world, is the idea of the "con." These scammers are working to gain your trust in order to convince you to bring them into your confidence (hence the term) and to make you believe that their sob stories or their threats or their bribes are true. As the old saying goes: "If the story sounds too good to be true, it probably is."

Before you react right off the bat—whether you're doing so out of charity, fear, or a desire to get in on the riches—take a moment to pause, think it over, and spend a few minutes to do some research. More often than not, you'll discover that it's just another con and shouldn't be trusted.

PROTECT YOUR PRIVACY ONLINE

MORE AND MORE OF YOUR PRIVATE LIFE IS AVAILABLE ONLINE EACH DAY. YOUR WORK CONNECTIONS, YOUR SOCIAL MEDIA PROFILE, AND THE PHOTOS ON YOUR PHONE ARE IN THE CLOUD, MAKING YOUR LIFE AN OPEN BOOK TO CRIMINALS.

As high-speed internet connections become available around the world, more and more of our lives are migrating online. People keep their résumés on LinkedIn, tweet links to their Instagram feed, and use Facebook for pretty much everything. And those photos and videos that used to eat up your hard drive space? You stashed those online, right? After all, if everything is password protected, it must be secure! That's the promise of "the cloud," a fluffy name for a network of servers on the public internet that let you stash your private documents, photos, and more. Imagine a massive train station with multiple banks of lockers. Anyone can enter the station, but if you stash your valuables in a locker, only you have the key that can open it—until someone secretly duplicates your key (i.e., steals your password) or just pries it open with a crowbar (i.e., uses malicious code that compromises your private files).

The cloud is growing every day—and not just with private files. Massive companies such as Amazon, Microsoft, Google, and others are migrating from earthbound data centers into cloud systems, too. In other words, a lot of private data is going into a public space. If the last two chapters have taught you nothing else, it should be that this trend is like catnip for cybercriminals.

NOTHING TO HIDE One of my friends is fond of saying, "Unless you called the police, don't talk to the police." He happens to be a thirty-year veteran police commissioner who probably knows what he's talking about. Why is this relevant here? Because the idea that "I haven't done anything wrong, so I don't need to worry about being hacked" is about as naive as the thought that "maybe if I explain to the officer that those drugs weren't mine, he'll let me go." So, why should you lock down your Facebook profile when all you post is pictures of your cat? Because that open-book page is easily hijacked. The next thing you know, Mrs. Whiskerson is wanted by Interpol for money laundering. Or a hacker using your name and email is asking all of your friends for a $100 USD loan. Guard your social media and other online accounts as carefully as you would other information.

FROM RUSSIA, WITH SPAM So assuming that the Russians, in some form or another, hacked the 2016 election in the United States, how did they do it? The size and complexity of the scheme is still being discovered, but here's one piece of it. Democratic candidate Hilary Clinton's campaign website was likely attacked by a Russian-linked criminal group using a targeted spearphishing barrage designed to look like it came from the Clinton campaign. The campaign's email system was breached, and those emails went out to her supporters. A whole lot of people who received those bogus emails clicked on them without a second thought. Each click got the hackers more access and information until they were able to access the campaign runners' accounts. The breach damaged the Clinton campaign multiple times before election day.

THE TRUTH IS OUT THERE As an increasing amount of our data is stored online, and our everyday lives unfurl in public, personal privacy and reputation come under threat in a number of new ways. And that means you need to update your strategies for staying safe. In earlier chapters, we've talked about broad-spectrum operations that seek to steal as much data as possible in the hopes that something will prove useful. In character attacks, the intention is often more personal and targeted, with the goal of damaging a specific person or group's reputation. These antisocial urges are nothing new, but technology makes it much easier to act on them. Back in the old days, people sought movie stars' racy photos through bribery or theft, but that was time consuming and expensive. And starting a nasty rumor? Sure, you could gossip, but how far would those lies really go? Today's troublemakers have more tools at their disposal that work anonymously—but you can still protect yourself and fight back. First, let's look at how they find your secrets.

UP IN THE CLOUD What we call "the cloud" is really just a bunch of computers sitting in data centers around the world talking to each other through global networks. These days, folks at a new start-up are likely to spend less time thinking about how many servers they need and more on how many cloud-computing resources they can use instead. The cloud provider takes care of all the hardware, software, security, and physical assets through these data centers, and it also assumes much of the risk of owning lots of technology. By relying on such infrastructure giants as Microsoft Azure, Citrix, Oracle, Google, and others to provide the basic infrastructure, as well as tens of thousands of companies to handle all the details, that new business can start up faster and focus on what it does best, as opposed to managing expensive, space-hungry server farms. Sounds awesome, right?

SAFETY CONCERNS As more and more companies go completely cloud-based, new vulnerabilities arise. While cloud providers are very careful about protecting their own resources from being hacked and destroyed, they are less able to influence what happens once data shifts into areas controlled by their customers. So, for example, it would be extremely difficult to successfully attack Amazon and gain control of its servers. But once that data has been dispatched to, say, an individual user's Dropbox account, it gets a lot easier.

We don't believe for one second that Netflix, or Amazon Video, or Gmail, or Dropbox are inherently insecure. But they rely on users, and users make mistakes. In fact, most hacks begin when someone makes a mistake. Picture a real-life delivery system: No matter how well the U.S. Postal Service protects your deliveries, once the mail is in your mailbox, it's your responsibility. If it gets stolen, you can't blame the letter carrier.

Remember: Even though John Podesta's weak password (which was "Fluffy1," by the way) may have played a role in the spearphishing attack on the Clinton campaign's official email, the attack would never have succeeded without a whole bunch of people absentmindedly clicking on an unfamiliar link.

This chapter talks about how to avoid those mistakes, what can happen if you don't, and how to clean up any resulting mess.

ANATOMY OF A CLOUD What we loosely refer to as the cloud is an ever-evolving collection of hardware and software—the servers that make up the infrastructure and the platforms and applications that let end-users access it.

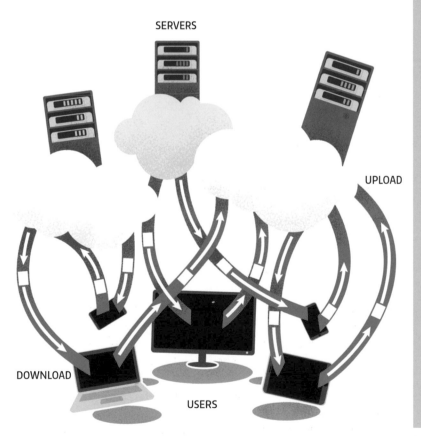

SERVERS

UPLOAD

DOWNLOAD

USERS

REMEMBER TO TURN ON PRIVATE OR INCOGNITO BROWSING TO KEEP SITES FROM STORING COOKIES ON YOUR COMPUTER AND COLLECTING INFORMATION ABOUT YOUR VISIT WITHOUT YOUR PERMISSION.

ONLINE ATTACKS, REAL-WORLD IMPACT

I'M A WOMAN WHO SPENT SEVENTEEN YEARS FIGHTING TO EARN MY TITLE OF VICE PRESIDENT IN THE OIL AND GAS INDUSTRY. IN A FEW SIMPLE KEYSTROKES, IT WAS ALL TAKEN AWAY FROM ME.

I used to think the internet was fun: posting updates about my life on Facebook, creating a LinkedIn profile to network professionally, tweeting random thoughts, and taking pictures of the world around me to put on Instagram—nothing prolific, just little things. Friends gave me advice about how online profiles could help my career in a future in which people would read about me online instead of talking to me in person. I trusted that the internet was a safe place to be.

Wow, how wrong I was!

I was thirty-seven years old when someone began to harass and cyberstalk me online. It began with false reports about me personally and professionally over numerous sites, from Twitter to Google+. The stalker created fake profiles of me on escort sites, harassed me on social media, and threatened me over the phone. Once, I was even sent a used condom in the mail, along with a note.

Then the cyberstalker raised the stakes and began to attack my friends, my family, and my company. When I didn't comply with the demands made of me, my tormentor posted bogus rip-off reports and reviews of the company I worked so hard to build. Every part of my life was targeted.

Until this began, I had never really understood the power of the internet or given very much thought on how I could navigate it safely.

As the cyberstalking intensified, as more information about me was posted in more places, I felt increasingly alone and that the rest of the world doubted me before even meeting me. I would walk into business meetings and be asked right away about intimate things no one would mention in the company of their own children, but because it was online, it was considered fair game.

People believe what they read online. Despite my efforts to set the record straight, I ended up losing contracts and ultimately my job. I couldn't trust anyone around me. I was under attack on all fronts. Some even exploited the situation to pressure me for money, blaming me for the impact my stalker had on their lives.

> ## "FOR THEM, IT WAS A SICK GAME; FOR ME, IT WAS REALITY."

Two years of relentless psychological terrorism left me feeling hopeless, helpless, and powerless. I had been completely violated. I had nowhere to turn, since all of my attempts to involve the FBI and local police were met with the same answer: "We don't have the resources to help with a situation that doesn't involve

murder." All I wanted was the answer to a simple question: "Why me?" Why would a stranger have so much hatred and feel the need to destroy a hardworking woman?

My now-husband and I had just started dating, and so it seemed likely that the attacks started out as an attempt by some unknown person to break us up. Instead, it forged us in fire. We were both broken to our cores, but we found our true love. We were married in the middle of this merciless attack, and now I have a teammate who is at my side until the end.

Two years on, the attacks still continue. I was advised that if I keep a low profile the attacker would eventually lose interest, but so far that has not been the case. So this year, I decided that I'd had enough. I decided to create a blog outlining all that had happened to me and the tools I found useful.

I am making sure my voice is heard. There are very few places to turn, and many are scams that cannot help you. My personal blog, www.stalkerexposed.com, explores in-depth the harsh realities of what can happen when someone wants to hurt you online. It is meant to serve as a reminder to everyone to take action and be safe online.
—Amanda Nickerson

LESSON LEARNED

Amanda Nickerson believes that the best protection online is to have a good password and two-factor authentication (see page 29) on every site you can. As a proactive measure, bolster your legitimate online presence and keep it up to date. This is huge: The less there is online about you, the easier it is for trolls and stalkers to make your life difficult. Laws lag far behind modern tech, and many of the companies that host mean and outright made-up content on blogs don't even respond to complaints or demands. You'll end up having to hire lawyers to do takedowns.

Spend your energy on genuine and meaningful content about what you truly do instead of sinking time into fighting lies. Google rewards solid content with better rankings. It takes time, but your peace of mind and career will benefit from a concerted effort to curate a solid body of online content about yourself and your interests.

CHECK YOURSELF You might not be able to stop someone from spreading fake stories about you, but you can make the harasser less likely to be taken seriously or even seen in search results.

Start by seeing what your online presence looks like right now. Using Chrome, open an incognito window or use duckduckgo.com and search for your name within double quotes ("Nick Selby"). The results will give you a sense of what a stranger Googling you would see. How does it look? Would you hire this person? Sell this individual a house? Go on a date with them?

Every month, check again to see what pops up. If your first check reveals nothing unusual, this exercise is just a formality. If, however, you discover that someone is trying to make you look bad, step up efforts to generate accurate content. Over time, the real you should rise in the rankings, while illegitimate sites fall away.

WATCH OUT FOR TROLLS Unfortunately, there's no shortage of women on the internet who still have to face random and sometimes extremely vicious harassment for little or no discernible reason or cause. While we fervently hope that in a few years our admonitions will seem as quaint and antiquated as a warning about spotting a dishonest footman, right now we'd be remiss not to touch on this unsavory topic. Women working in traditionally male-centric fields, such as gaming or technology, probably face the largest amount of abuse, but trolls can sometimes fixate on the strangest of things. One freelance journalist was testing blogging tools in order to set up a site; she posted a single goofy article on why she loves broccoli before abandoning the blog. Yet even this one article somehow touched a nerve: An unhinged stalker found that single post and made her life hell for more than a year. He made rape and death threats (the standard currency of the sexist troll), PhotoShopped her head onto pornographic images and mailed them to her employers and family, doxxed her, and even showed up outside her apartment to intimidate her in person. The threats never rose to a level that could get law enforcement involved, and it took her years to undo the damage.

She still writes under a pseudonym and is very cautious about using any social media. Sound like a one-in-a-million crazy story? Not if you read Amanda Nickerson's case (see pages 50–51). In an even more unnerving case, best-selling writer Jessica Valenti recently dropped off all social media after the commenters who regularly threatened her with rape and murder started making the very same threats against the writer's five-year-old daughter.

WHY THEY DO IT Popular writer Lindy West was plagued by a troll who got under her skin by creating a Twitter account in the persona of her recently deceased father to pepper her with insults and threats in his name. She wrote a fascinating piece about how painful this was—and unexpectedly got an email from the man behind the account. The resulting conversation (which you can hear on the popular podcast This American Life) was both illuminating and ultimately frustrating. He said that when he started harassing her, he was filled with self-loathing and was infuriated that she, a self-described fat woman, could be happy and successful. Why did this inspire him to torment and harass her? He had no good answer. It just seemed like the thing to do.

DON'T BE DISCOURAGED So, what's the takeaway for the average reader? Despite all of the above, the odds of this kind of random, sustained harassment are low. And, counterintuitively, while raising your profile may attract trolls, it will also give the kind of robust, impressive online presence that makes it more difficult for them to harm you. If you are harassed, report and block as necessary, and don't let them scare you away. In the unlikely event that such activity escalates, use the strategies outlined in chapter 11 to fight back.

REPUTATION MATTERS Modern commerce means doing business with all kinds of people you've never met but whom you still need to be able to trust. That's where online reputation comes in. Just as you have a reputation in your community, your school, your family, and with your friends, you also have one online that is based on your browsing history and activities.

The concept of online reputation was pioneered by the auction site eBay. As a global marketplace connecting buyers and sellers, the company had to offer tools to assure users that the strangers they are buying from are trustworthy.

If you use eBay, your reputation is based on whether you communicate well, pay on time, and send what was ordered . . . or whether you tend to stiff buyers or raise hell over trivial matters. On Uber, your ratings are those assigned to you by drivers after each ride. Airbnb users rate your home online, and you, in turn, rate their performance as guests. Right now, reputation is not transferrable—eBay users don't have access to your Amazon rankings, and Uber drivers can't see what Lyft thinks of you—but that might well change in the future as the concept develops. Reputation is perhaps even more important to small businesses, and we'll discuss these factors in detail in chapter 7.

SECURITY BASIC

EYES AND EARS In a 2016 photograph depicting Facebook CEO Mark Zuckerberg sitting at a desk, security folks noted that there was a piece of masking tape over the camera of his laptop. To those who wonder whether Zuck was being a little paranoid, he wasn't. In fact, closer examination showed that he had also disabled the microphone.

It was back in 2007 when I saw the first demonstration of a remote hack that stealthily turned on a user's camera and microphone, made a video, and sent it someplace, all without alerting the user. And the technology has really advanced since then.

To be safe, cover web-enabled cameras and microphones with masking or duct tape until you want to use them. This is what security nerds call a *positive security model*—that is, "deny by default, and allow by exception."

> **WARDRIVING IS THE "SPORT" OF SCANNING FOR UNSECURED, EASILY HACKED HOME NETWORKS. TO PROTECT YOURS, CHANGE THE DEFAULT PASSWORDS AND NAMES, AND ENCRYPT ALL TRAFFIC ON YOUR NETWORK.**

SNATCHING SECRETS FROM THE AIR Hackers love Wi-Fi, because these networks form one of the weakest points in an average user's online activity. And once they've breached your Wi-Fi, they can do a lot more than download Netflix on your bandwidth. A hacker can track and hijack the data you send and receive and use your connection to commit any number of crimes that could then be traced back to you. I've been on cases where the police literally kicked down a door, guns at the ready, to bust a major child pornography operation . . . only to find a very scared and confused older couple whose system had been hijacked. In that case, they were lucky the responding officers knew enough about cybercrime to suss out the situation. Not everyone is so lucky. Read on to learn some of the most common risks you face when going online wirelessly and how to defeat them.

At Home Loads of home networks are completely unsecured, which means they don't require a password for access. This exposes everything you transmit over that Wi-Fi connection to potential interception, and if that sounds like spy stuff, it shouldn't. You can learn how to harvest this bounty of information, if you are so inclined, with free software and instructional YouTube videos. Some networks are password protected, but the typical home user usually retains the default password that came with their wireless router. If this describes you, you may not be shocked to learn that there are entire websites dedicated to cataloging the default passwords for nearly every router ever made. Secure your home network with a strong password and change it often to increase security.

In Public Hackers love coffee shops and hotel lobbies. Harried travelers and groggy commuters constantly use free public Wi-Fi connections with no thought for safety. If the networks are unsecured, they can be "sniffed" just like your home network. If they are secured, hackers may set up a second network that isn't with a deceptive name. For example, search for Wi-Fi on your phone the next time you're sitting in the lobby of a large hotel. You may well see a long list—some of them belonging to nearby residences or businesses. But in addition to the official Whitby Arms Inn, there may be networks named things like "Guest Rooms," "Hotel Network," or "Lobby Internet." They may even be cleverly named to be listed alphabetically above the real network and therefore easy to select. Always ask for the name of the business's official network, and if you have the choice of a password-protected option, take it. It might even be worth paying a modest usage fee for enhanced security.

SURE SIGNS YOU'VE BEEN HACKED Despite your best efforts at keeping your network locked down, there's always a chance that a black-hat hacker has broken into it. Luckily, there are plenty of ways to tell if that's happened. Here's a list of potential symptoms to diagnose a compromised network. (There are plenty of other possibilities out there, too; if something just doesn't feel right about your network, dig deeper and you might find something as the result of a hack.)

Missed Connections If your network is running slowly during an apparent quiet time, it could be the result of someone else using your bandwidth. Too many connections from too many users can clog a network, especially a smaller one. Check and see who's logged on, and make sure the devices belong to people you know and trust.

A Lack of Control Are you unable to log on to the network? That may indicate that the login or password has been changed; a sure sign someone else has gotten in and locked you out. Be sure you change your network's default password at the very least.

Taking a Drive If your machine's hard drive is running slower than usual, and you notice the activity light flashing a lot more than it should, look into the situation a little further: Your antivirus software could be running a scan—or someone could have broken in and used malware to scan your disk looking for interesting data to steal.

Shields Down Is your antivirus software disabled even though you swear it was set to start every time your machine boots up? Or even though you swear you just restarted it five minutes ago? A malware infection can often disable antivirus software.

Unexpected Wares Your browser window didn't have that toolbar the last time you used it. And those pop-up windows weren't authorized either. What program just started during bootup? You didn't set that up, did you? Check to see if any extra software has been installed that you didn't put in yourself. Chances are it's the result of a hack.

No Shutdown The system won't shut down when you tell it to? You could be prevented from doing so by a hacker who wants to stay on the system. (But you can always pull the plug.)

T/F

WI-FI HACKING IS SO EASY, A KID CAN DO IT

TRUE You don't have to be trained in information technology work, or even be an adult, to know how to break in to someone else's system. All you really need is a computer with internet capability and a surprisingly short amount of time.

This was aptly proven during an ethical hacking demo in London in 2015, when seven-year-old Betsy Davies was shown a free YouTube video tutorial on how to fake a public Wi-Fi hot spot, then used it to get access to volunteers' computers—and in no more than eleven minutes.

In short, anyone could be a hacker—even a kid on a laptop in a nearby library or coffee shop. And a skilled hacker with the means and the intent can do a lot more damage than a curious kid.

- - - - - - - - -

A TANGLED WEB OF CONNECTIONS

I'M AN INVESTIGATOR WHO HAS SPENT MORE THAN A DECADE FIGHTING CYBERCRIME. IT'S COMMON FOR CRIMINALS AND CON MEN TO CONSTRUCT FAKE IDENTITIES AND COZY UP TO PERSONS OF INTEREST ONLINE. OF COURSE, THESE TECHNIQUES MIGHT ALSO BE USED BY COPS. HERE'S HOW A TYPICAL ONLINE "FRIEND" IS CONSTRUCTED.

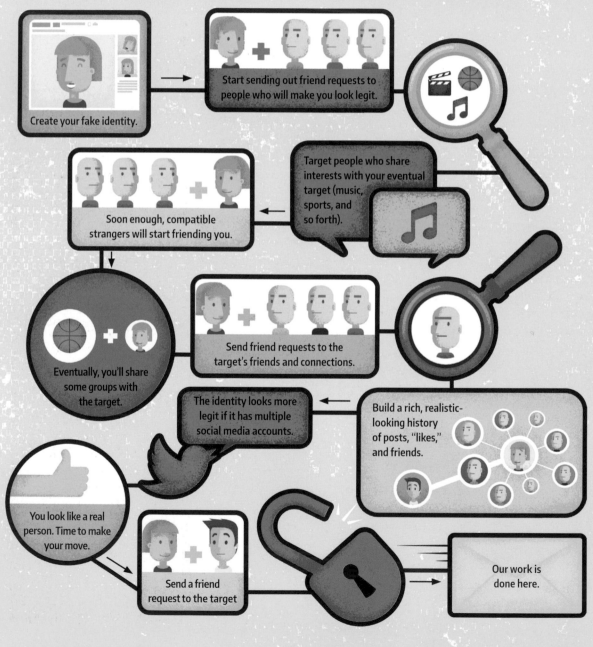

Create your fake identity.

Start sending out friend requests to people who will make you look legit.

Target people who share interests with your eventual target (music, sports, and so forth).

Soon enough, compatible strangers will start friending you.

Eventually, you'll share some groups with the target.

Send friend requests to the target's friends and connections.

Build a rich, realistic-looking history of posts, "likes," and friends.

The identity looks more legit if it has multiple social media accounts.

You look like a real person. Time to make your move.

Send a friend request to the target

Our work is done here.

THE PROBLEM OF OVERSHARING Lots of parents think their kids share too much info online, and they're right; we'll talk more about that in the next chapter. But while adults may be less likely to suffer physical harm, bullying, or ridicule than their offspring, they're often just as guilty of sharing too much with their friends on social media—and the personal, financial, and career consequences can be significant. Consider just a few cases of grown-ups not practicing what they (hopefully) preach.

Watch Me Fly The CEO of a major gaming company spoke publicly about battling a series of hacks from an adversary called Lizard Squad. Sometime later, he posted online about a trip he was about to take. The hackers identified the specific flight he would be on and forced it to divert by tweeting a bomb threat to the airline. If the executive hadn't provided sufficient data for the hackers to figure out his travel details, his business trip wouldn't have been disrupted.

Expensive Tweets Michael Dell, the tech executive, pays millions of dollars a year for security to protect his family from potential kidnappers and other dangers. Finding that his teenage daughter had tweeted links to photos of a family trip, with details on where they'd be for the next few weeks, was probably a little frustrating. As evidenced by the swift disappearance of her Twitter account.

Feeling Indiscreet The Israeli army was forced to cancel a military operation after one of the soldiers taking part in it posted the location and date of their planned attack on Facebook.

Busted! A decorated member of the UK's Buckingham Palace Guard was sacked after ranting on Facebook, calling Kate Middleton, the Duchess of Cambridge, a variety of inappropriate names. Back in the more mundane world, internet meme sites are filled with screengrabs of Facebook sequences like this:

 JOHN JAMES

Working for Luigi's sucks! I'm gonna steal all the breadsticks before I finish my next shift.

👍 💬

LUIGI'S PIZZERIA You can't steal breadsticks if you don't have a shift to come in for, John. We're letting you go.

LET'S BE FRIENDS In 2009, a male security researcher and white-hat hacker decided to see how easy it would be to get access to national security materials. He created LinkedIn and Facebook profiles for Robin Sage, a nonexistent female security expert supposedly working in "cyber threat analysis" at the Naval Network Warfare Command in Norfolk, Virginia, backed up by a fake work history and photos of an attractive young woman. In roughly eight weeks, the "Robin Sage Experiment" had access to emails, bank accounts, even classified military information from soldiers in Afghanistan. (I even broke my own rule in accepting the LinkedIn request. She had more than 250 connections in common with me, all from the InfoSec and intelligence community, and hey, she was cute. Luckily only my ego was damaged. —Nick Selby)

BUGS IN THE HUMAN HARDWARE

Security consultant and former hacker Kevin Mitnick has said that it's often far easier to talk someone into giving you their password than to try to break into a system yourself. That's where social engineering comes into the picture.

While not necessarily a con in and of itself, social engineering can be part of a con scheme (which we already covered in detail in the previous chapter). Either way, it involves manipulating a person into giving up sensitive information or performing actions that allow the "engineer" to acquire the information: an account login, email, password, credit card, or other sensitive material.

Remember that most businesses do not ask clients for their personal information, and most social engineering attempts are never even made in person. It comes down to trust . . . so, do you trust who's emailing you or calling you?

DRAWING UNWANTED ATTENTION

In 2011, our friend Aaron Barr, a man with years of intelligence and security experience, was a senior executive of HBGary Federal, selling advanced threat analytics and information tools to the government. Encouraged to put himself "out there" at conferences and in the press, Barr made a critical misstep: He said publicly that he had already identified many in the internet hacking community known as Anonymous. The result involved a social engineering attack against HBGary's computers that succeeded in breaching the email account of the firm's president and disseminating all the company's work and private communications to the public.

"I immediately wanted to fight to set the record straight," Aaron said. "I would even point journalists to specific leaked emails when they got their story wrong. More than one journalist said that he didn't have time to thoroughly research every story. In many cases, they go with whatever is hot and just get out what they can in the time they have."

Barr's side of the business went bust over the hubbub. The more he spoke out, the more dangerous it got—there were threats against his life and the lives of his loved ones. He and his family had to move multiple times. Ultimately, after a period of a few years of lying low and reestablishing himself, Aaron Barr has relocated, and he has begun to work again in an area in which he is particularly gifted: finding people on the internet.

Aaron believes that he should have laid low earlier and said less to stoke the flames. "I'm not the kind of guy to back down from a fight," he says, "but in this case, there were too many people, with too many motives to make me look bad. Shutting up and lying low was the best thing I could have done."

Tempting as it may be to share information online, stop and think at least once before you hit "post." Probably twice. Ask yourself, would it be safer to keep this closer to my vest? Am I likely to end up as the target of internet pranksters who may have no other goal than to prove themselves smarter than me (in their own mind, at least)? Would a major revelation help to fight cybercrime or just drive them further underground, spoiling for a fight? There are no one-size-fits-all answers, but you should take the time to ask the questions.

THE TAKEAWAY

Here's how to apply the lessons of this chapter, whether you're looking for basic safeguards, enhanced security, or super-spy measures to safeguard your privacy.

BASIC SECURITY

- Set all social media privacy settings as high as possible.
- Password-protect home Wi-Fi and encrypt with WPA2—never WEP.
- Don't accept friend requests from strangers you have not met personally.

ADVANCED MEASURES

- Only use the internet in incognito mode.
- Google yourself regularly and check what's said.
- Never use public Wi-Fi without a VPN.
- Restrict what you share on social media.

TINFOIL-HAT BRIGADE

- Keep nothing unencrypted in the cloud.
- Cover all computer webcams and microphones with electrical tape.
- Change user names frequently.

SECURITY BASIC

IS IT SAFE? Cloud storage gives us the promise that we can safely securely store any data we wish, and retrieve it at will thereafter—especially useful for small business owners. But the cloud is still ultimately a series of storage drives as part of a server in a location far away from your own—which means that any data stored there is likewise at a distance. If you lose connection with that cloud storage, or if the server goes down due to a hack or power outage, or if your account is compromised, you also lose access to all that data you've stored.

Basic data files, like simple documents, are mostly safe. But think twice about storing anything sensitive in the cloud, such as personal identification info, tax documents, or intimate photos and videos. Only store something if you're comfortable risking losing access to it or having it published somewhere online.

KEEP KIDS SAFE ONLINE

ONE OF THE BIGGEST CONCERNS PARENTS HAVE IS HOW TO KEEP THEIR KIDS SAFE FROM ONLINE PREDATORS, CYBERBULLIES, AND OTHER CREEPS. SAVVY PARENTS ALSO KNOW THEY SHOULD BE KEEPING THEIR COMPUTERS SAFE FROM THE KIDS.

Ask parents the fastest way to make their kid's eyes roll is and they'll tell you it's making the kid watch Mom or Dad interacting with technology. The generation of parents who grew up without smartphones, iPads, or the internet lack credibility with their children when trying to warn them about online dangers. If Mom can't figure out iMessage, how could she possibly know anything about internet safety?

This matters, because there are real dangers out there that your kids will almost certainly be exposed to—from online predators to annoying viruses that can destroy your data or encrypt your photos until you pay a ransom. Not to mention your kid clicking a video link that results in a spam gang using your router for distributed denial-of-service attacks against Walmart.

Kids born after the internet—sometimes known by us oldies as "digital natives"—grew up with this stuff and think nothing of digging into system preferences and settings we might not even know about. Which, in turn, means they can probably circumvent little inconveniences like that parental control software you installed.

The good news is that protecting a kid online has almost nothing to do with software and everything to do with straight talk.

IT ONLY TAKES TWO CLICKS In a New Zealand study on internet safety, researchers gave children between the ages of one and fourteen free access to a computer with unprotected internet access and told them to look for whatever they wanted online. Those kids ended up exposing the computer they were using to a virus within their first two clicks. No matter which topic they searched for, those two clicks likely exposed them to as many as twenty ads, most of them leading to high-risk sites offering pornography, gambling, get-rich-quick schemes, and the like. These kinds of sites are responsible for as much as 96 percent of the malware that's used by cybercriminals to access your machine in order to steal your information for various criminal enterprises. Learning to protect your kids and your machines go hand in hand.

The internet can create a false sense of privacy and community, with one-to-one conversations and anonymous screen names giving the illusion of a private space. Now, combine this with the natural tendency of teens everywhere to exaggerate, overshare, and test their boundaries, and you've got a recipe for all kinds of trouble.

CHECKING IN, CHECKING UP In 2016, 60 percent of parents with teenagers told the Pew Research Center that they'd checked out their teens' social media profiles and looked at the websites they visit, at least occasionally. And a similar percentage have friended or followed their teens on social media. About half of parents look at their teenager's phone call records or text messages, and know the passwords to their kids' phones.

Not surprising, younger parents are better at this—those under forty-five are much more likely to check up on teens and to check for the right stuff. For example, although every parent is worried about predators, younger parents are more likely to seek evidence of kids texting with unfamiliar friends and to look at call records and text records to ensure their kids aren't having inappropriate-sounding conversations with friends or strangers.

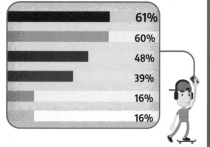

MONITORING YOUR TEEN
A majority of parents when have teenagers aged 13 to 17 have monitored their kids' web use. Here's how many do what:

Check browsing history	61%
Check social media	60%
Check teen's phone	48%
Block, filter, or monitor browsing	39%
Use parental controls on phone	16%
Use phone to track teen's location	16%

BAD DECISIONS All too often, teens are their own worst enemies. For example, a fair amount of "child pornography" actually originates with the teens themselves and then falls into the wrong hands. A kid may take nude selfies to share with a crush, assuming the images will stay private. Which they don't. The recipient might forward or share the pictures with friends, or might post the images as revenge after a breakup. And even if no one intentionally shares the pics, phones get stolen, iCloud accounts are hacked, and USB drives go missing.

The results can be tragic. Too many kids have tried suicide after such photos went public, often after vicious bullying. In some jurisdictions, teens have been prosecuted and branded for life as sex offenders for having nude photos of *themselves* on their phones.

Time for Real Talk This isn't a conversation any parent or child wants to have. But you need to, and keeping it fact-based is the key. Discuss real-life stories in the news, strategizing with your kids about how they'd handle these issues. Make it clear that many of the people hurt by online bullies or trolls didn't do anything wrong, but they still suffered.

KEY CONCEPT

STRATEGIC THINKING The specific online dangers change and evolve, but the tools kids need to combat them remain the same. A warning that stalkers may follow users on Instagram, or data on which online game has the most bullies, will date fast. Instead, encourage critical thinking by identifying the challenges and determining ways to avoid them. And always, *always* make it clear that you're there for them and will have their back.

BULLYING AND MOCKING Encourage kids to log, but not respond to, online bullying. If the bully persists, your child should feel comfortable coming to you for support. You should report it to school or police authorities.

REPUTATION ATTACKS Remind your kids that any private chat they have or image they share could go public. If they wouldn't want it posted on Facebook or mailed to Grandma, tell them to think twice before they hit "send."

PROFILING AND GROOMING If a new online friend suddenly starts commenting on your kid's Instagram, liking their Facebook posts, and retweeting them, be wary. This is often a precursor to asking for sexy pictures or more. Whether it's a cool-seeming stranger or a friend from school, teach your children to be careful.

TRUE STORY

CLOWNING AROUND

In 2016, the United States. and the UK were gripped by what came to be known as the "clown scare." In multiple locations, several people claimed to have seen "evil clowns" menacing them. Schools received messages from social media accounts in which people dressed as clowns threatened to blow up schools or instigate shootings. Some schools closed for a day or more, went on high alert after receiving such threats. But once law enforcement got involved, it all unraveled pretty quickly. Officers looked at each clown's social media account, and determined who their first followers were. This gave the police names of the kids who made the profile, as well as their close friends. Those kids were shocked when, the day after they made an "anonymous" threat, police officers swooped in to make arrests for a number of felony-level charges.

MORE THAN 60 PERCENT OF TEENS SAY THAT THEIR TWEETS ARE PUBLIC. PREDATORS AND CRIMINALS CAN HARVEST A LOT OF INFORMATION FROM TWITTER, SO BE SURE YOUR KIDS LOCK DOWN THEIR SOCIAL MEDIA ACCOUNTS.

SORRY, NO KILLER APP Every marketing claim made by parental-control apps is more or less true, but there's no magic bullet. Indeed, many parents have found that the most rigorous parental control software is more annoying than it's worth. Sure, that nannyware will keep your teenager from looking for topless models—and in some cases from writing a research paper about breast cancer or watching an ancient clip of Tom Hanks in *Bosom Buddies*. A smarter strategy involves securing your network and computers to a reasonable degree and then having more of those honest conversations with your kids about online safety.

Secure Your Network Your first step is to establish control of your Domain Name Servers (DNS). The DNS is the basic lookup tool used by everything on the internet to map a name to an IP address (which is a series of numbers, such as 208.67.220.220). We recommend setting your routers and devices to map to a DNS security site such as OpenDNS, which offers some basic free services that can help you control the sites your kids can visit, as well as premium options that allow for more customization. Your router may provide other parental control features as well, so check the specs online.

Secure Your Computers Remember, you're not protecting the computer from your kids. You're protecting your kids from danger. (Okay, fine, you're also protecting the computer from your kids.) Either way, you should password-protect your BIOS on Windows and Linux machines, or your firmware on a Mac, to prevent loading into bootable operating systems. If this sounds like gibberish, don't worry.

Google the key terms and you'll find helpful step-by-step instructions (as well as helpful instructions for your kids on how to confound you—read those too!). Give your kids nonadministrative and highly locked-down user accounts so that they cannot install software or make changes without your approval. And, of course, choose a strong, almost-impossible-to-guess password (see page 28).

Secure Your Kids This is the most likely point of failure. Remember, even if you successfully keep your children from accessing any "adult" sites, they can still be cyberbullied or stalked online by predators who know how to "groom" kids by empathizing with how mean and controlling their parents are. So, honesty is the best policy. These conversations should be age appropriate but specific: "There are bad people out there, and even though it feels as if you're anonymous on the internet, you're not." The point is not to scare them but rather to make them understand that the threats are real—and this isn't just you being overprotective.

SECURITY BASIC

MASTER OF YOUR DOMAIN You'd like to be able to trust that your kids are staying safe while using the internet, but just in case you need to keep an eye on them, there are ways. A home-based domain system (with its own URL) allows you to link multiple computers together and access information on a shared network, while bypassing your internet service provider to save on time and bandwidth. This whole mini-network can be protected from outside access with a firewall—and, properly configured, can also keep a log of any traffic going out to the internet, such as your kids' browsing history. The logs can record the time and date of the access, which device on the network it was done with, and the proper HTTP address and domain name, so you know where and when they're doing their browsing and with which mobile device or computer on your network. Plenty of companies offer services and software to build a home DNS, and there is no shortage of tutorials online that detail how to create your own.

GOOD TO KNOW

TAKE CONTROL When looking for parental-control software for your household, you'll find a range of options. Here are features we think are most important.

Be Inclusive Cover all platforms—Windows, Mac, iOS, and Android.

Set a Curfew Look for the ability to set times when the internet won't be available to the kids.

Stay in Control Look for remote management, monitoring, and control through a mobile app.

Know Now Set real-time alerts to text or email you if anyone tries to access a blocked site or search certain keywords.

Get Social Get your kid's social media logins and install software that alerts you to words or phrases and/or sends you random screen-grabs. Promise your kid you won't abuse your access, and keep that promise faithfully.

AGE-APPROPRIATE SURVEILLANCE When we use the word "surveillance," it's enough to make some parents cringe. Is that kind of thing really necessary? It sounds so . . . Orwellian. And what about letting kids make their own mistakes? Well, that's a nice idea when it comes to riding a bicycle or playing ice hockey, but on the internet it's possible to make mistakes that come at a serious cost. Kids have literally had their lives ruined as a result of something that started as harmless (to the kid) pranks. This isn't us being overly dramatic. Teens who send nude photos can end up on a sexual predator registry for life, which greatly restricts the jobs they can have, where they can live, and more.

We believe that you really do have to surveil your kids, if only to keep track of their promises to you. This chapter gives you a basic tool kit. For how to apply it, the chart below is a good start. You're probably not monitoring your three-year-old's phone calls or keeping up to date on the permissions for the Curious George game your sixteen-year-old is still registered for. But other nuances may be a bit trickier.

CHILD'S AGE	INTERNET GUIDELINES	WHAT TO WATCH FOR	ANYTHING ELSE?
0–2	· No internet access. Instead, download educational games.	· Anything you didn't load yourself.	· Be aware that toddlers learn how to operate devices quickly so don't have anything on your phone or tablet that you don't want them to open.
3–4	· Install positive security controls (which allow you to spell out what to access; everything else is off limits). · Activate Google Safe Search. · Allow downloaded and single-player online educational games only.	· Anything you didn't load yourself. · Chat requests.	· Be sure you're acting as a positive role model for kids who are just getting started using the internet on their own.
5–7	· Practice shared online time. · Utilize kid-safe search engines.	· Read all emails. · Read chat content. · Be sure you recognize all chat partners. · Make sure nothing's getting through the Safe Search settings.	· Introduce kids to the idea of cyberbullying, and discuss how to protect against mean kids online. · Be sure to explain why oversharing is bad, while keeping the discussion age appropriate.

CHILD'S AGE	INTERNET GUIDELINES	WHAT TO WATCH FOR	ANYTHING ELSE?
8–10	· Limit online time. · Set an audit trail through software or a router. · Limit social media. · Check social media interactions on all devices.	· Check your server and DNS logs regularly for inappropriate content or activity.	· Start talking about basic operational security (what not to reveal online, how to tell if someone might be a bad person, and so on).
11–13	· Do more consistent social media monitoring. · Set up Google keyword search alerts.	· Porn, gambling, meme, or image-sharing sites (which can have inappropriate content). · Pop-ups and adware. · Third-party toolbars and helpers. · Chat software.	· Time to have that awkward talk about adult content! · Keep talking about online security measures in more complex terms.
14–16	· Monitor in-app purchases, set limits as necessary. · Monitor mobile hot spots. · Check on chat software; make sure you know every app your teen is using.	· Fake accounts. · Duplicate accounts. · USB-based OS. · Watch for "sneaking" of computer access. Check time logged on versus how much you actually see your kid using the computer.	· Begin conversations about family responsibility, such as protecting the house from theft. · Make sure your kid knows how to spot online predators. · Start talking about college applications and what your kid's social media profile conveys to those schools.
17+	· Check texts and IMs occasionally for inappropriate images or messages. · By now, you should have established a respectful, trusting relationship. Good job! But don't slack off until that kid is actually an adult.	· Malware activations. · Dual-boot or bootable OS sessions. · Virtual sessions. · TrueCrypt-style steganography. · Check browsing history: Your router or ISP may have DNS logs that differ from your browser's history (which can be scrubbed).	· Keep the conversations going; be sure to praise good behavior. · Review your teen's online footprint together; play the part of a college admissions officer or potential employer. · Do occasional Google searches of your kid's name.

GOOD TO KNOW

FIGHTING BLACKMAIL

One cruel trick played by predators is charming underage kids into sending racy pictures of themselves or into undressing on a Skype call. Predators then use these images to blackmail the teen, asking for money or more-explicit photos and performances, threatening to send the original images to the kid's parents, school, or contacts list.

Reading this book puts you ahead of less-informed parents. Still, even smart, internet-savvy adults fall for online scams, and some creeps are frighteningly good at sweet-talking their way into a kid's confidences. Do your best to protect and inform your kids, and let them know that they can come to you if they get in trouble. Many kids are more scared of parents' anger or disappointment than they are of the blackmailer, and the results can be heartbreaking.

CASE STUDY

THAT AWKWARD TALK In a recent study, researchers found that 78 percent of high school students had watched porn, beginning around the age of fourteen. That means that even if your kid hasn't looked for X-rated content on the internet, their friends may well be sharing jokes and memes about adult topics, often with no idea what they really mean. I'm not here to give you parenting advice, but I will share my story. As a cybercop, I probably realized what my son was looking at more quickly than the average parent. In the pages that follow, I'll talk about how we handled the tech aspects. That's the easy part.

As a father, I realized that I couldn't stop my kid from exploring the internet's back alleys, but I didn't want him to get some wrong ideas. And that led to a very awkward talk in which, while he fidgeted and blushed, I told him that I wasn't going to police his viewing. But I did want to be sure he knew to look at those images as movies, not real life. I told him that the performers are actors, and the scenes tell you as much about how real couples act as Star Wars does about the space program. I warned him that he might see stuff that is scary or gross, posted for shock value, and that I wouldn't get angry or disgusted if he wanted to ask me questions about things he saw. Time will tell, but I hope and trust I've done my part to raise a boy who knows the difference between X-rated fantasies and real human relationships.

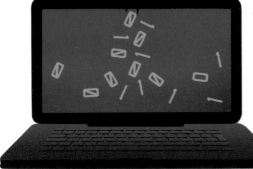

AN HONEST APPROACH Kids will watch or download things you don't approve of. This is just reality. Every parent has a different threshold for what that might be, and that needs to be the beginning of a family discussion. It's important to realize that while most warnings focus on porn, particularly the more disturbing images a kid might stumble upon, it's not the only inappropriate stuff out there. Spend a few hours in some of the least savory corners of Reddit or 4chan, and you'll realize how hard it is to shield a child from the darker side of humanity. The solution, as we keep reiterating, is to have those honest conversations . . . and back them up with technological solutions.

DAD, IT'S BROKEN AGAIN I already discussed how I used my son's teenage surfing as a way to have a parenting moment. Every situation is different, but I think it always makes sense to explain to your kids why you don't want them looking at certain sites. Otherwise, you're just mean old dad or mom never letting them have any fun. Start out by talking about how criminals attract us with sexy pictures, promises of free games or movies, get-rich-quick schemes, and more. The more a link begs you to click on it, the less likely it is that it will deliver on its promise. So, point one, the cake is a lie. Explain to your kids that they shouldn't download that pirated expansion pack—not because stealing is wrong (although you can remind them that it is) but because that freebie is almost certainly infected with all kinds of computer-destroying malware.

My son managed to completely trash his computer more than once before it occurred to me to make him a deal: I would not only restore his machine (again!), I would back off on monitoring him (a little) if he installed VirtualBox on his computer and configured it to spin up a virtual instance of Windows that he could use as a one-time computer. Then he could watch whatever he wanted . . . but if we got one more virus, he'd lose computer privileges for six months.

It took him three days to figure out VirtualBox, and we've been virus-free ever since.

KEY CONCEPT

RANSOMWARE
Remember what we said earlier about the infecting your computer in as little as two clicks? Sometimes the software that gets downloaded to your computer doesn't just damage it or use it for someone else's purposes but actually hijacks your PC, locking you out or encrypting your files until you pay for a "software removal tool" or give in to outright extortion. Some "warnings" you might see on your screen are normally meant to trick you into accepting the ransomware. If you see such a pop-up message, immediately close the window or reboot to stop it from taking over. In a neat new twist, hackers have started ransoming other web-enabled devices, such as smart TVs. If it happens to you, try a hard restart to factory settings and, if that fails, contact the manufacturer. Some are claiming that this only happens if you download pirated materials, but the jury is still out.

HIDING BEHIND THE NET When my mother asked me to look over an email she'd gotten from her internet pen pal, I routed it immediately to my internal Raised Eyebrows Department. Don't get me wrong, I'm not saying Mom isn't a catch, but the way she described the burgeoning romance was enough to make me suspicious. She met the guy through a dating site and began to chat, then flirt, then send emails. And after a little while, he professed his love for her.

Grooming Behaviors The general pattern here is the same, whether people are targeting kids or adults and whether they're seeking sex or money. What's referred to as "grooming" is all about finding common ground with a likely target, gaining the person's trust, and then going in for the kill. With Mom's sweetie, all I had to do was select a particularly poetic-sounding sentence, cut and paste it into a Google search, and bingo. There was the same message from the same guy, in thousands of posts.

Rule One If an offer sounds too good to be true, I can pretty much guarantee that it absolutely is. If someone online promises your kids exactly what they want, teach them to be careful. They should ask, "Who does this person claim to be? Can this person prove it?" If the stranger is legit, this should not be difficult. Otherwise, instruct your teens to keep their antennas up.

Warning Signs Anyone who asks your child for photos right away is suspect. As is a new person who always seems to know when your kid logs in to certain applications. Or if the stranger mentions having money trouble. Or if your child suddenly starts getting sexy pictures via email or text or Instagram—or any other way. Remember rule one. With your teen, try Googling a selection of text from the suspect emails—lots of these scammers work off of scripts and use the same emails over and over. Finally, if your teen asks to meet up and the person has a convoluted story about how they travel on business all the time, you can be certain that your kid is dealing with a fraud.

Con men use the internet's cloak of anonymity to steal money and maybe a heart or two. But online anonymity can mask an even more destructive face—that of the bully.

> TEACH YOUR KIDS TO BE GOOD INTERNET CITIZENS. EDUCATE THEM ON AVOIDING ANY ACTIONS THAT MIGHT HARM THEMSELVES OR OTHERS OR BREAK THE LAW.

STAND UP TO CYBERBULLIES Bullying can be more serious than almost any other online offense. And all too often, victims are told, "Just ignore them." That was bad advice in the 1950s when the bullies were waiting to beat you up after school, and it's even worse advice when the bullying can come from multiple sources, online and off. Kids commit suicide every year because of bullying, and even adults have their lives turned upside down. (If you haven't read Amanda Nickerson's story on pages 50–51, please do.)

If you or your children are targeted, contact authorities and do not accept no for an answer. Unfortunately, the internet can be rife with packs of bullies. We'll talk more about this in chapter 11.

THE TAKEAWAY

Here's how to apply the lessons of this chapter to help keep your family safe from cyberbullies, online predators, and pesky malware..

BASIC SECURITY

- Monitor all social media accounts your child uses.
- Talk to kids about what's safe to share.

ADVANCED MEASURES

- Restrict and lock down your home network.
- Log traffic and use software to track net activity.
- Restrict social media sharing.
- Install GPS tracking apps on kids' phones.

TINFOIL-HAT BRIGADE

- Lock down all social media accounts to private.
- Use spyware to track all online activity.
- Use a private LAN for kids' computers and aggressively blacklist sites and categories at the router.

T/F

CYBER-BULLYING IS KIDS' STUFF

FALSE Cyberstalking, cyberbullying, and online smear campaigns are a growing and increasingly destructive form of abuse. It affects young and old, male and female, and even the famous: Jennifer Garner, Ellen Page, Ciara, 50 Cent, and a long list of other celebrities, have been cyberbullied and stalked or had private photos stolen and circulated. Comedian Leslie Jones took a break from Twitter after unrelenting racist and sexist abuse that began when she appeared in the remake of *Ghostbusters*. Social media sites have taken steps to stem the tide, but the most important step is to realize that anyone can be a victim.

THE INTERNET OF THINGS

THESE DAYS IT SEEMS LIKE EVERY GADGET, APPLIANCE, AND ACCESSORY IS "SMART" IN SOME WAY. THE INTERNET OF THINGS GIVES MORE POWER THAN EVER TO US AND OUR OBJECTS . . . AS WELL AS TO HACKERS AND SPIES.

A few years ago, the idea of a web-enabled clothes dryer or "smart" light bulb sounded like either marketing hype or really boring science fiction. Now, every new day seems to bring a new object that's been made "smarter." Sometimes the integration is subtle and virtually seamless—we've gone from watches to smart watches or Fitbits. Elsewhere, an upgrade solves a real problem in a helpful way. For example, video-enabled doorbells let you see who's at your door via a smartphone app, no matter where you are.

Internet-connected objects offer new powers and more control over our lives—who wouldn't want to turn off the stove from any room in the house, have the garage door open as you pull into the driveway, and keep that pesky neighborhood cat out while letting your cat in?

So, what's the catch? Well, what personal data is each device collecting? How is it stored? Is it secure? Has it ever been hacked? Do I need to update my password? (Pro tip: Companies are notoriously secretive about being hacked, for good reason.)

Smart objects are designed to make our lives better, to give us more control. But what if someone else takes control? For that matter, why hack a light bulb? Can a baby monitor be weaponized? What is my dryer saying about me to the fridge, anyway?

HOUSTON, WE HAVE BEVERAGES

The first internet-connected object, long before smart devices, was a humble vending machine. Back in 1982, computer science students at Carnegie Mellon found a solution to that age-old problem, the heartbreak of walking all the way to the soda machine only to find it's out of your favorite brand. That's fifteen minutes you'll never get back. Students Mike Kazar, David Nichols, John Zsarnay, and Ivor Durham installed microswitches in the vending machine and wrote a server program to report the machine's current status and temperature. That soda machine has become an internet legend and went on to inspire other projects, perhaps most notably the Trojan Room Coffee Pot at University of Cambridge, which, in 1991, used the world's first webcam so that students could see how much coffee was left in the pot without leaving their workstations.

PHONE IT IN When we think of the ways smartphones have changed our lives, we tend to think of the convenience of texting, checking email, booking rides, and ordering takeout. We tend to forget that our phones are really small, powerful computers—and they're almost always online. BlackBerries and other internet-connected mobile devices before the iPhone existed were almost exclusively used for business. That changed in 2008 with the iPhone and its app store—and the countless apps developers could easily make for nearly everything. That drove demand for always-on internet and ubiquitous wireless access. Wi-Fi is often cheap or free and available mostly everywhere, and internet access is considered a basic human need on a par with water, heat, and electricity. In this world, it only makes sense that we'd control our homes, pets, and cars by phone.

A BRAVE NEW WORLD At its most utopian, IoT promises a futuristic home, with connected appliances you can control from your phone, thermostats that let you wake up to the perfect temperature, and light bulbs that turn themselves on in the morning and off at night. You can turn on the air conditioner in your smart car before you leave the beach so that the interior is refreshingly cool by the time you're ready to drive home. With a verbal command, devices in your home can play your favorite song, tell you the weather, or cue up a movie. Here are some of the "things" you might encounter.

Smarter Homes One of the richest sources of smart devices, homes have a staggering number of options for technological conveniences, and it increases every year—from smart coffee machines that will brew coffee when you wake up (based on a signal from your Fitbit) to internet-connected refrigerators, washing machines, dryers, and ovens. There's even an internet-connected sous vide that enables you to start that slow-cooking dinner from miles away. Pop your iPad into the docking station on your intelligent yoga mat, and you have a personal yoga teacher, without having to venture out of your home.

Smarter Cars Your home isn't the only thing getting smarter. If you've purchased a new car in the past few years, you know that its bells and whistles are all connected to the internet. Cars today come with their own internet connections and smartphone apps. Forgot where you parked? No worries, your smartwatch remembers!

Smarter Pets Most pets are microchipped—this makes it easy to reconnect with them if they get lost. But the "internet of pets" doesn't stop there! Radio-frequency identification (RFID) and microchip-activated pet doors only allow the pets you select to come and go. The problem with tag-related access is that pets can lose these tags, and they can be costly to replace. Got a pet that is overeating? The internet of pets can help with that, too, with smart feeding bowls.

Smarter You There are now plenty of wearable fitness trackers such as Fitbits or FuelBands, or Misfit fitness and sleep trackers. But internet-connected devices don't stop there. Smart medical devices already exist, and they're only getting smarter—like pacemakers or insulin pumps that can share your medical data with your doctor. Of course, these developments aren't without their own problems (see page 84 for more).

HOW PREVALENT IS SMART TECHNOLOGY IN U.S. HOMES?

As the internet of things weaves its wires throughout our lives, so too do objects in our homes become fully connected—and interconnected.

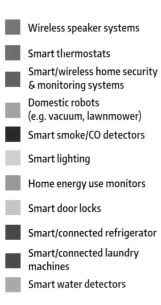

- Wireless speaker systems
- Smart thermostats
- Smart/wireless home security & monitoring systems
- Domestic robots (e.g. vacuum, lawnmower)
- Smart smoke/CO detectors
- Smart lighting
- Home energy use monitors
- Smart door locks
- Smart/connected refrigerator
- Smart/connected laundry machines
- Smart water detectors

> MANY PRODUCTS HAVE DEFAULT PASSWORDS. CHANGE THEM RIGHT AWAY TO DISSUADE HACKERS FROM NOT ONLY STEALING YOUR DATA BUT HIJACKING YOUR DEVICES AND USING THEM TO CAUSE TROUBLE.

COSTS OF CONNECTIVITY Who wouldn't want all this neat stuff? Besides the sometimes high price tags, what are the downsides?

Unsecured Data Much of this technology is still so new that the bugs are still being worked out . . . or surfacing unexpectedly. After all, Bluetooth and smartphone apps are great tools, but they were never designed to safeguard confidential information. As voice command becomes more common, most folks haven't yet wondered whether Alexa ever stops listening. If the microphones are always on, and the internet connection always live, what's happening to all that data? Is anybody listening? And if so, who are they and what are they capable of? There are lots of questions, and the answers are only just now being made public.

Listening Ears In today's competitive business environment, the "consumer data use revenue model" is a big sell point for any digital entrepreneur looking for funding to launch a new product or company. In other words, most new trackers and sensors have a plan to use the data they collect on you for other purposes—for example, to sell to marketers. Some have even more nefarious uses.

The Bottom Line The problem with filling our world with smart objects and sensors—although they will absolutely help us have better lives—is a data problem. We need to ask what data is being collected, how secure the storage servers are, how that data will be used, and who ultimately owns it. Right now, data collection, use, and ownership stays with the business that makes the product. Most businesses are motivated by money—and your data is money.

Then there are other problems with these devices. Recently, I wanted to turn on my Hue connected light bulbs. But before I could do so, I was forced to update my software—turning a swipe into a five-minute update! While a Nest thermostat is fantastic for remotely monitoring and controlling the heating and cooling in my home, hacks giving access to my data could make my home a target for thieves. The Nest could even be tampered with to spy on me!

These devices may be smart, but they don't have morals. Where most people would feel uncomfortable listening in on private conversations in private homes, these objects have no such ethical guidelines built into them. It's the people who work at the manufacturers who decide what is fine to listen in on, what data to store, how to store it, and what to do with it.

GOOD TO KNOW

WHAT A DOLL! Science fiction is chock-full of stories about kids' intelligent toys being used for good or evil. This concept has moved another step closer to reality with the coming wave of toys that use Bluetooth and a phone app to communicate with children. One of the first "smart" dolls was My Friend Cayla, who connects to the internet and relays information provided by a child, something like a cuter version of Siri with some parental controls. More sophisticated options include Hello Barbie, a model of the world-famous doll that uses pre-scripted lines to communicate with a child while also building a cloud-based bank of information to better tailor those conversations to that individual. Concerns about these toys include uncertainty about what information is stored, how it will be used, and how secure it is from theft. In addition, hackers have found that they're able to hijack the Bluetooth signal that controls Barbie from outside a home and "speak" whatever they please to kids in the doll's voice.

KEY CONCEPT

HIJACKED DEVICES
Your data isn't the only thing at risk if hackers break in to your devices. They can also use them to attack websites. These Distributed Denial of Service (DDoS) attacks happen when computers or other internet-linked devices are programmed to repeatedly request a specific website. The millions of requests from hacked devices overwhelm the server, causing the site to go down. In 2016, malware called Mirai (Japanese for "future") used IoT connectivity to launch a massive DDoS attack. It identified over sixty default user names and passwords and took over devices such as baby monitors, DVRs, and security cameras, in a network called a botnet. In October 2016, this botnet hit DNS service provider Dyn, as well as PayPal, Spotify, Wired, GitHub, Twitter, Reddit, Netflix, Airbnb, and others. At least 1.2 million IoT devices are possibly still infected by Mirai.

HOME SWEET CYBERHOME

Unless you have a love affair with high-tech gadgets and a salary to match, your home may not be quite this wired. That said, you may well have more connections to the internet of things than you realize. As IoT technology gets cheaper and more ubiquitous, more and more devices will be talking to each other.

PEOPLE
- Bluetooth headset
- Fitness tracker
- Smartphone
- Smartwatch
- Medical device
- Hearing aid
- Baby w/ smart diaper

BEDROOM
- Smart adult toys
- Smart mattress (sleep tracker)
- Intelligent yoga mat

KIDS ROOM
- Baby monitor
- Smart doll
- Nanny cam bear

LAUNDRY, UTILITY ROOM
- Washer
- Dryer
- Vacuum cleaner

GARAGE
- Vehicle (with GPS, stereo, Bluetooth phone connectivity)
- Garage door
- Smart lawnmower
- Water heater

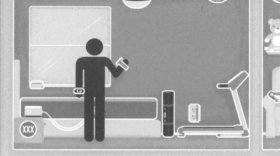

LIVING ROOM, STUDY, ETC
- Nest/thermostat/AC
- entertainment system/TV
- Personal computer
- Game console
- Smart TV
- Exercise equipment (elliptical, treadmill)
- Pet door (synced to pet's collar)

EVERYWHERE IN GENERAL
- CO & smoke detectors
- Wireless speakers
- Smart light bulbs
- Smart heating vents (open and close to shift heat to cooler areas)

BATHROOM
- Smart scale
- Smart mirror
- Toothbrush
- Smart breathometer (tells you if you have bad breath)

KITCHEN
- Fridge
- Coffee maker
- Sous vide
- Dishwasher
- Oven/range

EXTERIOR
- Lights
- Cameras
- Door/window alarms
- Sprinklers
- Front door (pet door too; see above, left)

BAD VIBRATIONS How personal is the data your devices might be sending to manufacturers, with or without your consent? The delightfully named field of teledildonics is exploring ways that sex toys can be controlled via a smartphone app or Wi-Fi. Some even involve webcams that allow partners to share supposedly private videos with each other. But, as with many other IoT devices, there are two big problems for users: Manufacturers tend to put little effort into security, and they may be gathering data without your knowledge. In 2016, hackers at DEF CON decided to hack a smart vibrator as a conversation starter about assault. They weren't surprised at how easy it was to take over the device, but what was unexpected and shocking was that it was secretly gathering data on the user's favorite settings and other factors and uploading them to the company.

GETTING SMARTER ABOUT SMART DEVICES "Smart object," "smart device," "internet-connected device"—these are all terms that are used to describe objects that connect to and use the internet. Some devices connect directly to the internet from your home network through your router, while others connect to the internet via an app on your smartphone. When these devices are connected to the internet, you're able to remotely control them—or program them to do things on their own. Say, for example, you want to schedule your lights to turn on at dusk, whether you're at home or not. Smart light bulbs become smarter when connected to a weather site that provides the time of sunset. There are websites with hundreds of these kinds of programs.

Making Smart Objects Even Smarter IFTTT (which stands for "if this, then that") is a free web-based service that allows you to create applets that enable you to connect your smart devices together. This presents an interesting range of security advantages but also concerns. The biggest concern is, what if someone managed to hack IFTTT? The way data is stored makes it unlikely that a breach would give hackers access to vast amounts of data or control of all your devices. Still, it is a worry for some. A more immediate concern is the tendency for oversharing. The more data about your movements you put out there, the more opportunities there are for someone to track you for nefarious purposes.

On the other hand, many IFTTT applets are specifically designed to make you, your home, and your family more secure. The table on the facing page depicts a number of these options to show you how versatile IFTTT can be. Do the security and convenience of these applets outweigh concerns about hackers? Only you can make that call, but the possibilities are intriguing.

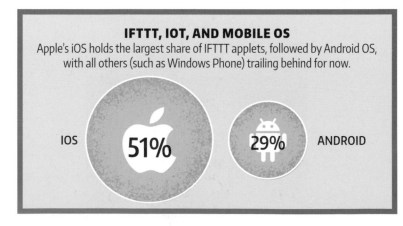

IFTTT, IOT, AND MOBILE OS
Apple's iOS holds the largest share of IFTTT applets, followed by Android OS, with all others (such as Windows Phone) trailing behind for now.

IOS 51%   29% ANDROID

IF THIS . . .		. . . THEN THAT
Security camera recognizes a face at the front door		Porch lights turn on
Vehicle approaches garage and Bluetooth system is recognized		Garage door opens
Pet (wearing smart collar) gets close to pet door		Pet door opens
Doorbell is rung		Security camera sends picture of visitor to smartphone
Last person leaves house		House confirms oven is off and sends that person a text
Sun rises (or sets)		Blinds on sun-facing side of house close (or open)
Smoke alarm goes off		Next-door neighbor receives text message on smartphone
No inhabitants are detected inside home		Security cameras and alarms activate
Wake-up alarm goes off in the morning		Coffee maker in kitchen turns on
A door or window is forced open		A security camera sends a photo and text to smartphone
An inhabitant in the house moves from one room to another		Air vents close in previous room and open in new room
No inhabitants are detected inside room		Roomba activates and begins cleaning the floor

GOOD TO KNOW

ACCESS GRANTED With all these concerns about data privacy, there is some good news: Plenty of companies are now working on technical solutions to help enable more control and security with data sharing. UMA, which stands for User-Managed Access, is an OAuth 2.0 protocol that defines how developers can enable a smart object to engage in secure selective data sharing. This protocol makes it easier for developers of software and hardware to let the owner of the smart device specify what data they would like to share and what to keep restricted. The use of UMA removes the security burden from the item's manufacturer and also gives consumers or owners more power over the proliferation of their own data. UMA is a protocol that can be used right now, in fact—we just need to get more manufacturers to use it in their products. Look for it when you buy, for greater safety.

CHECK EVERY DEVICE YOU OWN FOR PRIVACY SETTINGS AS OFTEN AS YOU CAN, AND SET THEM TO THE HIGHEST POSSIBLE RESTRICTION TO MINIMIZE WHO GETS TO SEE AND USE YOUR INFORMATION.

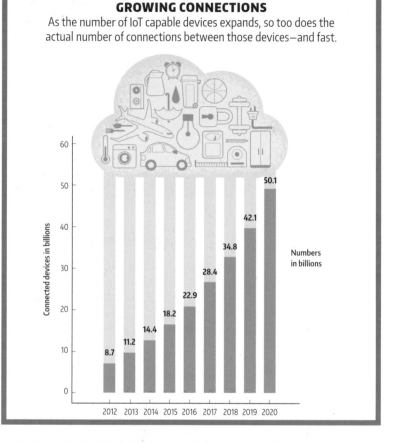

GROWING CONNECTIONS

As the number of IoT capable devices expands, so too does the actual number of connections between those devices—and fast.

Connected devices in billions

Numbers in billions

8.7 (2012)
11.2 (2013)
14.4 (2014)
18.2 (2015)
22.9 (2016)
28.4 (2017)
34.8 (2018)
42.1 (2019)
50.1 (2020)

UNDER SURVEILLANCE All around us, data is collected about our activities and behavior. From what route we took to get to the grocery store (Waze, Google Maps) to whom we're messaging (Facebook Messenger, WhatsApp), companies that build the software we use are constantly tracking and monitoring us. Much of it is used for what's called "surveillance marketing."

May Be Relevant to Your Interests Surveillance marketing happens when companies such as Google, Facebook, Amazon, and other sites observe the information (data) you generate by using their services. Google was a pioneer in this field with what was then called "contextual marketing" back in the early 2000s. After the release of Gmail, Google began monitoring the contents and context of messages in order to show advertising based on all of those messages. If you emailed your mom about an upcoming vacation to Bali, Google might show you ads about airfare specials, travel, or vacation activities in the South Pacific. After some outcry, the company put a halt to this particular practice. But it's still the case that if you search shopping sites for the perfect pair of rain boots you'll likely be stalked by advertisements for rain boots for the next year or so, across a range of sites.

Information, Please Data is collected about you not just on the Web but when also you use your phone and smart devices. Your smart thermostat, your car, your light bulbs, and your fitness tracker are spying on you and reporting back to . . . someone. Imagine the offline version of this—a company representative listening in on your private conversations and following you around to see what you buy You wouldn't stand for that kind of behavior in the real world, but it's become part of what you expect online.

Big Brother Wants to Watch The internet of things (especially IFTTT devices) works by monitoring and responding to everything you do. But this also means that someone else could be watching. In 2016, former FBI director of national intelligence James Clapper informed a Senate panel that the government had known about the potential to use IoT and IFTTT to spy on their users. Privacy advocates, in response, are encouraging consumers to use end-to-end encrypted smart devices and are pushing for more privacy laws.

KILLER APP

DITCH THAT TAIL If a creepy dude is following you, you can duck into a safe space to escape. But online spies aren't immediately visible, so how do you ditch them? If you want to minimize instances of being tracked online, use an ad blocker such as Adblock Plus and site tracking browser plug-ins such as Ghostery. But beware, some sites will intentionally offer less functionality or ban your browsing outright if you use an ad blocker. If you are concerned about Facebook using your personal data (or any apps that you have given access to your Facebook data), remove Facebook Messenger from your smartphone, and use a secure messaging system such as Signal, Wickr, or SMS. And if you want to stay truly stealthy, you can always use Tor (see page 170 for more).

T/F

A SMART APPLIANCE WAS WITNESS TO A MURDER

TRUE (kind of) All police knew in late 2015 was that Victor Collins and James Bates had spent an evening soaking in Bates's hot tub listening to music streamed by an Amazon Echo. According to Bates, he went to bed early. When he woke up in the morning, Collins had apparently drowned.

Police were suspicious, and it quickly became clear that Collins was in a fight before his death and had likely been drowned. But there was no witness. Or was there?

Investigators served a warrant to Amazon, hoping the Echo had recorded anything of interest. It's unlikely that Alexa will take the stand, but an interesting precedent may have been set. Siri, where's the best place to bury a body?

YOU CAN HACK A PACEMAKER

TRUE While this is true, it hasn't happened to a person—yet. Many medical devices have wireless functionality to share information with your doctor to see how well the device is working. The late hacker Barnaby Jack did pioneering work here, and in 2016, security firm MedSec hacked pacemakers and defibrillators and then licensed to a Wall St. hedge fund the data on how they did it. Medsec's hacks included sending a shock—wirelessly. The firm shorted the stock of the manufacturer, St. Jude Medical. St. Jude denied this, but the FDA and DHS confirmed the hacks. Barnaby's most famous hack had been of an insulin pump, causing delivery (in a lab) of a lethal dose of insulin.

ON THE ROAD As the internet of things grows, it's no surprise that it extends to affect our vehicles as well. Our cars, trucks, and SUVs are not only a source of more data for companies to mine but open drivers up to a new range of threats.

Digital Carjacking In 2015, *Wired* magazine writer Andy Greenberg volunteered to drive a Jeep Cherokee while hackers attempted to control it remotely. Hackers Charlie Miller and Chris Valasek were able to take control from ten miles away by laptop; Greenberg was helpless to stop the duo from controlling the A/C, radio, windshield wipers—and even stopping the transmission. The same sort of vulnerability had been demonstrated previously in 2013, with hackers accessing the brakes, horn, seat belt, and steering wheel of a Toyota Prius with Greenberg behind the wheel. Only in recent years have legislators begun to set electronic security standards for automobiles, and some auto makers have issued even recalls for their vehicles. Nonetheless, the risk of hackers assuming control to control it, stalk the driver, or steal relevant data still exists. So far, the hacks have only been done to help understand a car's vulnerabilities. However, it is no longer the stuff of bad movies to imagine that you could be hurtling down the freeway when, with no warning, your doors lock, brakes fail, steering freezes, and seat belt clicks open. But hey, you can still stream your Spotify playlist, so that's good.

Driverless Cars Anyone who has been paying attention knows that the era of the driverless car is finally upon us, as more and more companies follow in the steps of Google's extensive testing. Safety concerns are, of course, paramount, and there have been a number of fender benders (mainly the driverless car being rear-ended by a car with a driver inside), along with reports of cars blowing through red lights or stop signs. A Tesla in "autopilot" (semi-driverless) mode was involved in a fatal accident back in 2015, but after much investigation, the manufacturer was absolved of fault. Indeed, the statistics show that cars with autopilot are actually involved in 40 percent fewer incidents.

THE TAKEAWAY

As the internet of things is a relatively new phenomenon, ways of keeping yourself safe mainly involve doing your research and using common sense.

BASIC SECURITY

- Research purchases before you buy.
- Change your modem and router passwords to something other than the factory default.
- Use screen lock codes on all mobile devices.
- Isolate IoT apps.

ADVANCED MEASURES

- Ensure that medical devices are locked to only critical services.
- Ask device providers about wireless security.

TINFOIL-HAT BRIGADE

- Set up a separate home network with a separate firewall with all your IoT Devices behind the firewall.
- Place IoT devices on a virtual LAN segment.
- Install surveillance software to collect data packets sent from your devices through your network.

FLYING THE UNFRIENDLY SKIES
Not all the problems surrounding smart vehicles and devices are limited to the ones that are on the ground. Radio hackers have broken into American and British air traffic control and transmitted bogus flight information to pilots. These "ghost transmissions" have requested that pilots change their landing plans and diverge from flight paths. Up until now, the instances of air traffic hacking have been low in number; only twenty ghost transmissions have been properly identified and no one has ever been caught in the act, let alone prosecuted. The equipment for breaking into the pilot's signal sets a user back about $450 USD—still well within the means of a determined troublemaker. And the real problem is that there's currently no technology available to block the unauthorized people who are making these transmissions.

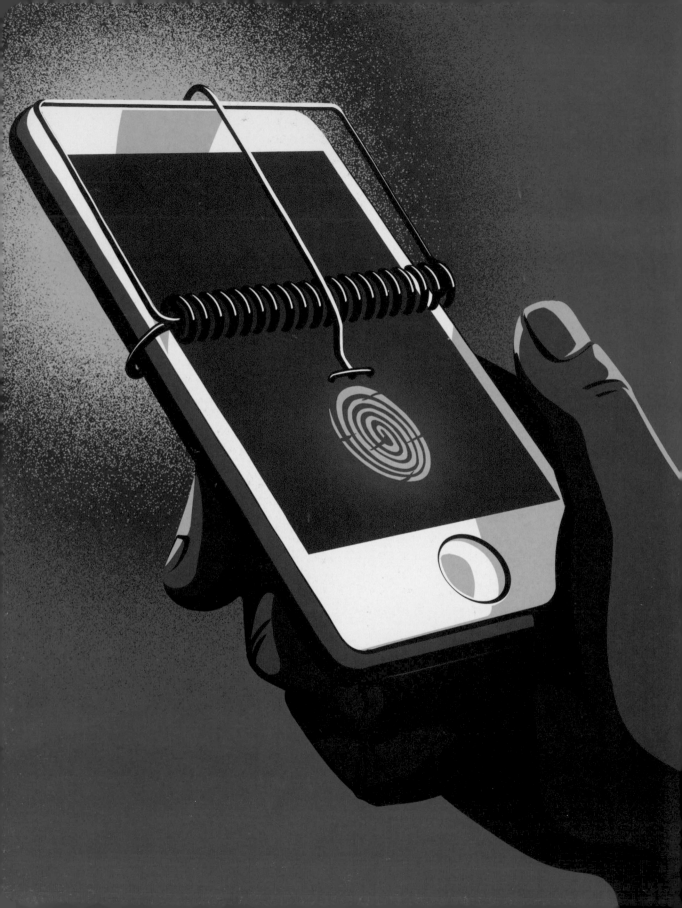

NOT JUST PHONING IT IN

NEVER MIND THAT YOUR SMARTPHONE IS MORE POWERFUL THAN THE COMPUTERS THAT POWERED THE SPACE SHUTTLE. THE REAL ISSUE WITH PHONES IS THAT THEY CONTAIN A COMPLETE SET OF THE METADATA OF YOUR LIFE.

The era of pocket-sized computers predicted since the 1950s began in earnest in 2007 with the release of the first iPhone. Today, people all over the planet carry with them at all times a device capable of accessing the internet, the Global Positioning System, and hundreds of millions of nearby devices. We've passed the point of inflection: More people these days view websites with their mobile phone than on desktop or laptop computers.

We entrust to our phones more data than we traditionally stored on our home computers. Our phones know exactly where we are and where we've been, and can tell exactly where we're going. They know how much we have in our bank account and what we've bought on Amazon; they can open our front doors and start our cars; they know how much dirt is on our floors and whether our smoke alarm batteries are charged. They can transmit and receive voice and text communications between us and anyone else in the world. They know, increasingly, whether we need to buy a quart of milk or how many people read that story we posted on Facebook. Golly, that's a lot of data about us. Say, these phones are secure, right?

Well, not really. In fact . . . if you're not very careful, your phone can cause you some serious problems.

PIKACHU, I CHOOSE YOU! You may have seen one futuristic thriller or another, wherein the bad guys use some clever or disgusting method to breach security using a victim's thumbprint. But as it turns out, you don't need to be a supervillain or a high-tech professional to be a cyberthief. In 2016, the *Wall Street Journal* reported the story of a six-year-old who used her sleeping mother's thumbprint to unlock the mom's phone. She then went on to order $250 USD worth of Pokémon toys from Amazon. When her parents received thirteen different order confirmations they assumed their account had been hacked—until their daughter proudly announced that she'd been shopping just like Mommy. Parents of precocious children might want to rethink their password strategies, wear gloves while napping, or just make sure their smartphones are kept safely out of reach.

THE BEST WAY TO ASSURE YOUR PRIVACY AND SECURITY IN CASE YOUR PHONE IS LOST OR STOLEN IS TO SET A GOOD PASSCODE OR SECURITY PATTERN—AS COMPLEX AS YOU CAN MAKE IT.

BASIC PHONE SETTINGS So, you've just gotten a new phone! Congratulations. As you set it up, you should keep some things in mind in order to make security a basic part of the way you operate your phone from the beginning, as opposed to trying to patch it up later.

Lock It Down First, enable your phone's "Lock SIM Card" option, which will require a password to access your SIM card every time that the phone is rebooted. This is in addition to your phone's screen lock. The SIM lock secures the network access card, while the screen lock protects the phone itself.

Secure Your Screen Next, enable your phone's screen lock to keep snoops, busybodies, and evildoers from going through your stuff. If you can use a password or passphrase (instead of a PIN), do it. If you use a PIN, make it six numbers or longer. Set the lock timer low enough to ensure that it automatically secures itself when you leave it in a restaurant or in a taxi, but long enough to avoid entering your passcode every three seconds—two minutes should be a good degree of compromise.

Don't Get Fingered Plenty of smartphone models offer fingerprint detection as a quick way of unlocking and accessing your phone, but despite its convenience, you should deactivate that setting completely and rely on PINs and passwords for your privacy and protection (see page 98 for more).

Stay Cryptic Deactivate location settings unless you specifically need them. Often you'll end up needing to turn location on for specific apps when you need it—say to get driving directions—but it's worth the (relatively minimal) trouble.

Use Only What You Need You should also turn off your phone's Wi-Fi, tethering, hot spot, Bluetooth, and near-field communication settings. All of these are useful, but you should only activate them as you need them rather than simply leaving them on by default.

Block Out Finally, consider blocking your phone's caller ID to maintain maximum privacy. The people you call will see you on their end as "Unknown" or "Private Caller" and as a result some of them will no doubt ignore your calls. This is a borderline "Advanced" or even "Tin Foil Hat Brigade" level of security, for those who really want to go stealth. You're going to be leaving a lot of voicemails, since most "private callers" are debt collectors, scammers, doctors' offices, and cops.

GOOD TO KNOW

In this chart, apps in tan are generally configured to protect your privacy. Items in yellow can be safe so long as you watch your settings and understand what you are trading for fun or convenience. Items in orange are known to have poor privacy controls and/or to sell user information to advertisers or the government.

APP OR FEATURE	EXAMPLES
SECURE MAIL	K-9, Proton, SafeMail for Gmail, Inky
PRIVACY APPS	Ghostery; Dash VPN; TorGuard VPN; Orfox: Tor Browser for Android; Orbot: Proxy with Tor; Red Onion
SECURE CHAT	Signal, WhatsApp, Wickr
TRANSPORTATION	Uber, Lyft
SOCIAL MEDIA	Facebook, Twitter, Instagram, Kik, Reddit, Vine, Tumblr
SPYING SOFTWARE	FlexiSPY, mSpy
GAMES	Fruit Ninja, Angry Birds, Despicable Me, Words with Friends
FLASHLIGHT	Flashlight, Brightest Flashlight

SECURITY BASIC

FIND MY PHONE

There have been, over the past few years, a number of high-profile cases in which users have located their stolen smartphones (both iOS and Android) or tablet through "find my phone" applications. These are generally a good idea to have running because as it turns out, petty or nonprofessional thieves are generally not that careful when it comes to using stolen phones—they just turn them on and use them. Duh.

When your phone is stolen, be sure to go to the police station and fill out a theft report. When you locate your stolen phone, you don't want the first conversation with the fuzz to be about your phone having been stolen. Rather, you want to say, "Hey! You know that phone I reported stolen in report MP1735068? I found it! It's at 356 Main Street." That has a better chance of success in terms of getting the cops to help you retrieve your device.

**WHAT'S IN YOUR
SMARTPHONE?** A wise man may have once claimed, "You are not the contents of your wallet," but that was before smartphones came along. Your wallet probably has a few pictures, some cash, and a range of cards: identification, insurance, debit, and credit. All of these are meaningful and should be replaced if your wallet is lost or stolen, but it doesn't compare at all to what your smartphone can contain: Hundreds of photographs and videos, hundreds or even thousands of emails, open apps with personal info such as banking, passwords to every online account you have, as well as continuous access to any and all further sensitive information sent to those accounts. In short, you are the contents of your phone to anyone who steals or breaks into it. Keep it even closer to you and more secure than your wallet.

YOUR PHONE IS NOT A WALLET . . . EXCEPT WHEN IT IS Your mobile phone is not inherently unsafe, but you will need to understand the relative risk that a given app presents and compare that to the reward you get by using it. When you think about using a new app, consider the following: What do I get, and what does it cost me, really?

It's the same as giving your personal information to supermarkets in exchange for a frequent shopper card. You enjoy savings, but the store maintains a complete dossier on what you buy—and don't. They know the days you go shopping and the amount they can count on you to spend, and they have a good idea about how to increase that amount. That data is theirs to do with as they please, and they often sell it to the highest bidder.

Banking on You When your bank asks you to try its new app, they obviously want you to use it because it teaches them more about you. What's the risk? The bank will always pay back your losses if the app gets breached. If you find it convenient to deposit checks from your phone or transfer money, that's a great deal. But you need to understand what you are giving them and the cost of the worst-case scenario. Is Google Wallet or Apple Pay as convenient as Google and Apple make it sound? It depends on your circumstances—if you live in a big city, sure; out in the country? Probably not.

What Do You Need? Do you have a nine-year-old who uses your phone for multiplayer online gaming, or is this a business phone with a few key applications and a good password? Everyone's needs are different, and one man's convenience is another man's big shiny target. Give some serious thought to what you want to get from your apps and how you use your devices. For example, use a different mobile device for sensitive apps like banking, and only use the device for that—this gives you much of the convenience but with less risk.

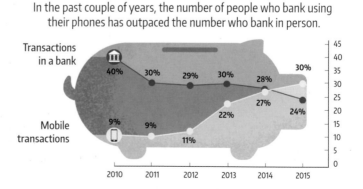

THIS LITTLE PIGGY WENT ONLINE
In the past couple of years, the number of people who bank using their phones has outpaced the number who bank in person.

Transactions in a bank

40%   30%   29%   30%   28%   30%

27%

22%   24%

Mobile transactions

9%   9%   11%

45
40
35
30
25
20
15
10
5
0

2010   2011   2012   2013   2014   2015

You probably have a pretty good grasp of your phone's basic features, but it may still be able to surprise you with what it can do—and how it can affect your data safety. Here are some possible security risks you might want to check out.

FEATURE	WHAT IT'S FOR	WHAT YOU SHOULD KNOW	WHAT TO DO
GPS	LOCATION AND DIRECTIONS	Your GPS provides your location to any app in the phone—including apps you might not know about. Some phones keep records of where you have been stored in a file on the phone and uploaded to the cloud . . . somewhere.	Be selective in granting permission to apps to use your location. Shut off location services unless you are actively using them for something.
WIFI	INTERNET CONNECTIVITY	You often need Wi-Fi to cut down on mobile data use, but Android phones use Wi-Fi to more exactly locate you. By sensing the relative strength of available Wi-Fi hot spots, they can triangulate your position much more closely than with GPS alone.	Keep Wi-Fi turned off unless you're using it; deny permission to "Use Wi-Fi to enhance your location."
CAMERA	SHOOTING PHOTOGRAPHS OR VIDEO	Your phone can add metadata to your photographs, including the location, time and date, and other personal information.	Look in your phone's settings to limit the metadata added to photos or EXIF data.
FACEBOOK	SOCIAL NETWORKING	Facebook takes about as many liberties with your personal data as anyone outside Fort Meade.	Only you can make the risk/benefit analysis as to whether Facebook's juice is worth the squeeze of handing Facebook all your data.
MICROPHONE	SPEAKING, RECORDING	Your phone's microphones (there are often more than one) are high-quality directional and omnidirectional mics capable of grabbing crisp, clear audio. Any hack that grabs your mic can use it as a very powerful listening device.	This is the hardest to mitigate. Tin Foil Hat Brigade members recommend removing your phone's mic entirely and using a lanyard mic such as the one that comes with your earbuds.
BANKING APP	MONEY MANAGEMENT	Banking apps can be very convenient. They have also been hacked, forcing data to be forwarded to places you might not want.	Be careful about using these tools, and never for business (banks don't have to reimburse business accounts after fraud).

T/F

POLICE NEED A WARRANT TO SEARCH MY PHONE

TRUE The United States Supreme Court in *Riley v. California* (573 U.S. ____ [2014]) found that police generally may not search digital information on a phone seized during an arrest without a search warrant. The exception for searches incident to arrest does not apply to these cases. The Fourth Amendment is the key here; the court's ruling stopped highly questionable practices—like using technology to dump the contents of phones on the side of the road. Understand your rights: If an officer (or TSA employee) asks, say you will be happy to comply with a search warrant. Even if you are under arrest, they can't just search your phone. Don't let them.

- - - - - - - - -

ERIC OLSON ON METADATA: The best way to think about metadata is that it is data describing other data. A movie review blurb like "A delightful yarn," could be considered to be metadata about a two-hour movie in which a clever and complicated story is told. Metadata is a high-level description or abstraction that gives you an understanding of the data itself. If I'm standing in a bookstore, and I've got to give my boss a report tomorrow, I'm wondering which of these books contains the content I need, and there may be eleven factors I'm considering: Does the book fit in my briefcase? Do I know the author? Are they a trusted source? Is this thing too heavy to hold up while I'm in the bathtub?

All those are factors about the book that have nothing to do with the content in the book. So, for each datum, there's the message or content of the datum itself, and there's a whole range of information around it that is about it but not that which is in it.

Listening In If we look at the metadata in the context of phones, we see that it can be incredibly powerful. The best known example is the government looking at metadata of your phone calls. If you receive a phone call from a doctor or a blood lab and speak for three minutes, then you call a pharmacy and speak for two minutes, and then you

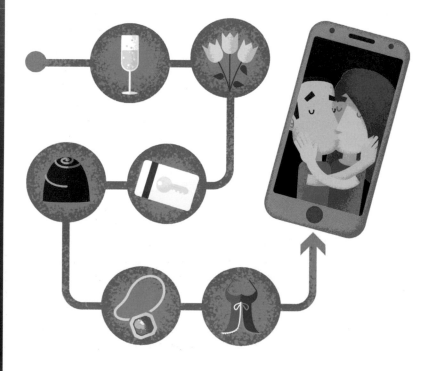

call an AIDS patient advocacy group and speak for nineteen minutes, the government might consider it likely that you've just gotten a diagnosis of HIV.

Similarly, if you should receive a call from a known drug dealer lasting for a minute or two, and then follow it up by making a set of calls to three of your friends and speaking with each for under two minutes, this might lead investigators to believe that you have just agreed to buy drugs and to share them with your friends.

Learning Your Habits Metadata needn't be about legal matters; it can describe lifestyle choices and behaviors. Imagine if your health insurance company had access to call metadata or the information from your supermarket's club card, so it could base your rates on how often you ordered Chinese takeout or how many frozen pizzas you bought—or whether or not you've renewed your gym membership? What do you think a private investigators would make of smartphone metadata showing a purchase of Champagne, followed by flowers, a three-minute call to the Squeezy Acres Motel, then online shopping with a purveyors of chocolate, jewelry, and sexy undies . . . and finally a series of texts to someone who is definitely not your wife?

Simple Summaries Metadata needn't be complicated; in fact, one single character can provide a wealth of information. For example, your grade of B in a class is metadata describing an entire semester of your work. It's long been used to identify kids in need of extra help in school. An entry of 1 instead of a 0 in a field like "gun owner" can change the way the police approach a house or car. Of course, such a simplistic on/off bit of metadata can occasionally be subject to error, especially when entered initially by a human user—consider the aforementioned example of whether the police believe you to be a gun owner when approaching your house or car.

It's All in How You Use It Despite all this, though, there are actually more good uses than bad of metadata—meta-analysis (the aggregation and analysis of metadata from multiple sources), for example, is probably the best way to understand the overall efficacy of a drug—have a look at the Cochrane Collaboration (http://www.cochrane.org/), and you can understand how examination and aggregation of the metadata around drug trials can help produce better health outcomes. Metadata, like technology, is a dual-use issue. Its "goodness" or "badness" depends on the context and your perspective. —Eric Olson

TRUE STORY

THE FBI VS. APPLE

Public debate over law enforcement access to phones came to a head after the FBI sought, under the All Writs Act, to force vendors to provide cops with a "backdoor" to encrypted iPhones. The case arose after the FBI sought possible evidence on the iPhone used by a mass shooter in San Bernardino who attacked a government office in December of 2015.

The FBI sought to force Apple to break the strong encryption of that device. Apple refused, arguing that a backdoor was the same as a master key for cops. The police argued that Apple was creating a safe space for criminals.

Finally, the FBI bought a tool from hackers to break the protection on the iPhone and get the data they sought. The argument is not yet over; both sides are still waiting for a great test case, ultimately to be settled by the United States Supreme Court.

SELLING YOUR SMARTPHONE'S SOUL

Some apps are notable for doing a lot more than you were expecting when you downloaded an installed them. A now-famous smartphone flashlight, for example, is the poster child for this kind of unwelcome surprise. The Brightest Flashlight app, estimated by the Federal Trade Commission to have been downloaded tens of millions of times, stole identifying data, location, calendar information, camera and microphone access, email, and network surveillance—essentially giving your whole phone to the app. The app makers were then selling all this information to advertisers. The lesson? If you decide to install an external app, examine the permissions that it asks for. Use apps that only ask for permissions related to their tasks. If an app is suspect, ask questions, or look for one that uses fewer permissions.

IS THAT JAMES BOND IN YOUR POCKET, OR ARE YOU JUST HAPPY TO SEE ME? When people with national security clearances enter Secure Compartmentalized Information Facilities (SCIF)—the secure rooms we have seen in the movies—all mobile phones are confiscated. If a phone is detected, the owner is in huge trouble. This is because of the fact that as phones get better, they offer really neat spy tools that many people forget they have handy at all times.

Utilize Electronic Eavesdropping Want to record a meeting? There's an app for that. Leave your phone on the table, with its screen turned off, and you can covertly get a high-fidelity stereo recording of what's being said. There are also free or low-cost apps that record all incoming an outbound calls, spoof your number to whatever you want it to be, detect mobile networks and cellular signals, and do several other cool tricks as well.

Keep Track of a Subject Would you like to spy on someone else's phone? If you have access to it (and flexible morals) you can easily download and install apps that can track, say, your significant other or your kids. If you have no morals whatsoever, you can place spyware on the phone of a coworker or your employer, for a little industrial espionage. Of course, this works both ways: Your employer can place spyware on your phone and probably be within the law in several states, so long as he or she owns the phone. Spyware can allow you to, for example, remotely activate the microphone on a phone and listen in on the conversation or access the GPS to both track location and set up a geo-fence, which will send you a text message when your quarry leaves a designated area.

Steal Secrets If you've set up spyware on someone else's phone (or someone has done it to your phone instead), you can also peek in on social media accounts, snatch the very keys they type with a keylogger, crack passwords, and access other wonderful features. You have a robust piece of covert surveillance gear that used to require a boxy suit and a mustache to twirl—or at the very least, nation-state funding.

Don't Just Make Calls We as users continue to get it wrong, seeing smartphones as "phones" and not "computers." They are decidedly the latter, and they're capable of doing all the things a computer can do—we should never to forget that.

KILLER APP

A number of apps can help you feel more secure, whether you fear for your civil liberties or your wallet. It's definitely a good idea to check out the iTunes or Google Play Store regularly to learn about new applications—just use the keywords "safety," "spyware," and the like. Read the reviews online to learn about possible vulnerabilities and to find the best functionality for your needs, and always check the permissions when installing and app.

Electronic Frontier Foundation The EFF's app updates you on civil liberty issues and helps you tweet to government leaders.

ACLU The American Civil Liberties Union lets you record police activity and upload it in real time to a cloud server to document police abuses or brutality.

bSafe Personal Safety This, and similar apps, can identify people in your network to help you stay safe, travel in groups, and get help.

Other Safety and Privacy Apps Many personal-safety apps are also available, both from cities (such as the BART safety app for riders of the San Francisco Bay Area transit system) as well as private concerns (such as the ICE—In Case of Emergency—app)

SECURITY BASIC

KEEPING A SECURE CONNECTION Your smartphone's Wi-Fi and Bluetooth capabilities mean you can connect anywhere, anytime, to any network or device. No question, the ability to work and play online using your smartphone is awfully convenient. But every single public access point you encounter is a chance for someone else to peek in and even steal sensitive information from you while you're online. And if your Wi-Fi is turned on, products like the WiFi Pineapple can trick your device into sharing information without your knowledge. To be safer when using your smartphone, laptop, or tablet, use only secured connections. Use a VPN connection, or just always skip free unsecured public Wi-Fi. Your phone still can use its own mobile data in plenty of places. It might cost a little more depending on your plan, but the added safety is often worth the money.

SECURITY BASIC

DUAL-PURPOSE TECHNOLOGY

The spyware stuff can be really great for keeping tabs on your kids or checking in on elderly relatives. But remember, if their phones or yours are hacked, that's a lot of data falling into the hands of strangers. In information security and intelligence operations this is known as dual-purpose technology. A classic example is the IMSI-catcher (mobile phone eavesdropping device) that has put Edward Snowden in a huff. You are truly happy that the government has it when using it, for example, to home in on a kidnapper or a child molester, or to locate mobile devices in prisons. But it's not so great when the government focuses attention on whatever you happen to be doing—especially if you don't want to explain why you were at the Shady Acres Motel the night before. That's a classic dual-purpose quandary.

THE NATION-STATE THREAT No discussion of surveillance and mobile phones would be complete without a mention of Edward Snowden (for more on Snowden, see page 182), who in 2013 stole a large number of files classified by the U.S. government as Secret, Top Secret, or higher. Regardless of whether you think he is a hero or traitor, he opened our eyes to the technical capabilities of well-funded actors, such as organized criminal gangs and nation-states.

In 2016, Snowden spoke on HBO about the issue, demonstrating how to remove components from a smartphone to remain completely safe (you can see this on YouTube). In VICE's documentary *State of Surveillance*, with Edward Snowden and Shane Smith, Snowden described why exactly you would want to remove the microphones and the cameras from your phone (note the plural tense of those components) and demonstrated just how to do it.

"Every part of private life today is on your phone," said Snowden. "They used to say that a man's home was his castle. Now, his phone is his castle."

Listening In Snowden was interested in IMSI-catchers, devices that, essentially, impersonate cell towers and intercept cell phones. Here's where we must warn that Snowden's understanding of how police use IMSI-catchers is absolutely flawed— use by U.S. police of IMSI-catchers is dramatically less common than Mr. Snowden believes. But he is correct in his assertion of the capability and ease of acquisition of an IMSI-catcher by police or private individuals for getting hold of the phone data you are sending. The solution is to follow security protocols appropriate for your risk profile, and follow them religiously.

Risk Assessment If you are an activist or protester or someone who takes on activists or protesters, you should consider the likelihood of attacks against your electronic life to be rather high. If you're a soccer mom, you should consider the likelihood of a targeted attack rather low—that doesn't mean that you don't face mobile threats, such as malware, spyware, and the like. It just tells you how to set your expectations of privacy and therefore your security posture.

Staying Safe The Electronic Frontier Foundation (EFF) provides a very good starting place for this exercise in an article on its website titles "An Introduction to Threat Modeling." To give you an idea of how this works, we authors use encryption for our email and hard drives, and, wherever possible, we use encrypted voice and text through the application Signal. The EFF recommends Signal specifically because of the strength of its cryptographic implementation, its ease of use, and its "zero-knowledge" model— even if Signal itself is hacked, it cannot turn over any of your messages because it cannot read them. That's a powerful feature.

Finally, as we keep reminding you, encrypt your phone, and use a strong password to protect its contents—do not use a fingerprint. The reason for that is that U.S. courts have ruled it legal for police to use your fingerprints to unencrypt something, but it is currently not legal for the police to force you to turn over your password.

THE SPY IN YOUR POCKET We reached out to our friend, J.D. LeaSure, to ask what he thought of the cellular spying world. J.D. owns ComSec LLC, which is in the business of technical surveillance countermeasures—you know, in the movies when they have a guy come in and sweep for bugs? That's J.D.

"One of the worst culprits of cellular spying involves a program named FlexiSPY," he said. Highlights include call interception, SMS interception (including WhatsApp and other popular "secure" SMS applications), SMS tracking, password cracking, a digital recorder that can be set remotely not to ring/vibrate or light up, etc., etc. It's one bad mofo! Oh, and it works on both Android and iOS," says J.D.

"Another one is mSpy, which can track call logs, GPS location, and metadata about how the person is living his or her life—from calendar updates and text messages to email and web history. It, too, works with both Android and iOS.

"The best bug in the world is right there in your pocket . . . and adversaries know that your phone is never more than six feet away, and you always have it with you, and turned on."

IF YOU DECIDE TO USE YOUR PHONE TO MAKE ANY SORT OF PURCHASES, PAY BILLS, OR DO BANKING, LOG OUT OF THE SITE IMMEDIATELY AFTERWARD, AND NEVER SAVE THE LOGIN INFO ON YOUR PHONE.

T/F

COPS HAVE A RIGHT TO TRACK YOUR PHONE

TRUE The case here is *United States v. Skinner*. The DEA tracked Mr. Skinner's vehicle by repeatedly "pinging" a phone it knew he had, as Skinner drove on public thoroughfares from New Mexico to Texas. When Skinner stopped for the night, the cops arrived and a drug-sniffing K-9 indicated the presence of drugs. Police searched; discovered the dope, the guns, and the phones; and Skinner and his son Samuel were arrested. In this case, the United States Court of Appeals for the Sixth Circuit ruled that "If a tool used to transport contraband gives off a signal that can be tracked for location, certainly the police can track the signal."

FINGERPRINT VERSUS PASSCODE In 2014, a test case made it to the Second Circuit Court in Virginia in *Commonwealth of Virginia v. David Charles Baust*. This is not a federal court but rather an important state court. Baust had been charged with trying to strangle his girlfriend. The state said that Baust and his girlfriend had set up video equipment in the bedroom, and the victim admitted that the video from that setup transmitted to and was saved on Baust's mobile phone (for exactly the reason you'd suspect). The state argued that the setup had captured the event on video, so it got a warrant for the phone. Baust refused to provide the fingerprint to unlock and decrypt the device, claiming that passcodes and fingerprints are protected by the Fifth Amendment, which prohibits forced self-incrimination. The state argued that the production of the credentials was not "testimonial," for the simple reason that it had already been established that the video was on the phone—the existence of the recording was a "foregone conclusion."

A Key Difference In his ruling, presiding judge Steven C. Frucci drew a very important distinction between using a passcode versus a fingerprint. It had been established in case law, the judge ruled, that a passcode represents testimonial communication. The ruling was determined because a passcode—unlike bodily measurements or fingerprints, or a voice exemplar or handwriting sample—is not a physical thing; instead, it is something that resides in a person's mind. Therefore, forcing the defendant to reveal his passcode would mean that the government is forcing him to testify and divulge potentially self-incriminating evidence.

On the other hand, the fingerprint is not considered testimony. Judge Frucci pointed to well-established concepts, stating that "There is a significant difference between the use of compulsion to extort communications from a defendant and compelling a person to engage in a conduct that may be incriminating. . . . The privilege offers no protection against compulsion to submit to fingerprinting, photography or measurements, to write or speak for identification, to appear in court, to stand, to assume a stance, to walk, or to make a particular gesture."

A Legal Matter Basically, all that fancy legalese means that the court can't force you to produce the combination to a lock, but it can make you produce the key to a lock.

As a result of this ruling, it is considered a pretty solid bet in America that the government can make you produce your fingerprint, but it can't force you to produce your password or passcode. Lesson? Don't use fingerprints or other biometric identifiers like retinal scans or handprints to protect your data. Always use a strong passcode.

THE TAKEAWAY

You use your phone for just about everything in your life, so there's no one-size-fits-all set of security protocols. Think carefully and act accordingly.

BASIC SECURITY	• Use a good password—never your fingerprint. A good password is more than six characters or numbers, or a good pattern. • Encrypt your phone. • Use a phone-locator app in case it is stolen.
ADVANCED MEASURES	• Limit the number of days of email that can download to the phone. • Use a VPN for browsing in public. • Use 2FA for all apps that you possibly can. • Limit location services and Wi-Fi use. • Ensure limited metadata is saved with images.
TINFOIL-HAT BRIGADE	• All of the above—and remove the cameras and microphones from your smartphone. • Use encrypted DNS (will only fix Wi-Fi). • Regularly reflash your phone to factory settings. • Limit data usage.

TO KEEP YOUR PHONE TOTALLY SECURE WHEN NOT IN USE, TURN IT OFF COMPLETELY— AND THEN REMOVE THE BATTERY IF POSSIBLE TO AVOID ANY PASSIVE DATA COLLECTION.

The story of the internet is about more than laptops and cell phones and smart cars. It's also about the people behind those devices, and the things they desire—wealth, power, connection, influence, and more. With just a few clicks of a mouse, we're able to shop online, look for dates, educate ourselves. On the other hand, there are no shortage of people employing the same connectivity to scam others, take their money, damage their businesses, destroy their confidence, break their hearts, and cause plenty of other intangible forms of damage—and they don't necessarily act on their own. Online anonymity is afforded to the larcenous and the amorous alike, and if an internet mob gets going, they can do just as much harm as any rioting mass of people in the public square. So how do you protect yourself, your business, your bank account, and even maybe your heart in this realm? In the next several chapters, we'll have a look at how to safeguard all of these things.

CYBER SECURITY AND YOUR BUSINESS

WHETHER YOU'RE SELLING HANDMADE CRAFTS OUT OF YOUR HOME OR RUNNING A BOOMING ONLINE EMPORIUM, YOU FACE A UNIQUE SET OF RISKS IN THE NEW GLOBAL MARKETPLACE AND NEED TO PROTECT YOURSELF AND YOUR CUSTOMERS.

Starting a business is incredibly risky and incredibly expensive, and keeping one going is rarely much easier. Unexpected costs are everywhere, and it goes without saying that business owners will try to save money wherever they can without sacrificing the quality of their goods. Over and over we see that it's very tempting to scrimp on computer costs.

Before we talk about software licenses and antivirus programs and password strength and monitoring, we need to ask a basic question: What, exactly, qualifies as a business these days? Two of the contributors to this book run businesses from their homes. They have business bank accounts, bookkeepers, and accountants, and they do all their work while sitting in a comfy chair, wearing a pair of bunny slippers. The internet has changed the economy so much from the way it was even in the early 1990s that sometimes Boomers, Gen Xers, and Millennials have vastly different concepts of what constitutes a business.

If you remember what a Beanie Baby is, you remember the turning point in America after which people could actually make a living selling junk on the internet. Today, twenty-five million people make their living in whole or in part by selling things on auction sites.

THIS IS SERIOUS BUSINESS Profit is revenue minus expenses, and one thing home businesses do is generate revenues that are typically much higher than profits. A business phone account can cost twice as much as an ordinary one, for example, and reserving space for a business is yet another overhead cost. There's also the cost of handling utilities and licenses, the expenditures in hiring and training employees if you expand, maintaining the various supplies needed for the business's upkeep, insurance fees, and loss of inventory from accident or theft—and that's to say nothing of the cost of any electronic or internet-based aspects. All these expenses mean that a loss can be devastating to the owner. And when home businesspeople, for example, conduct business on the same poorly protected Wi-Fi network on which their teenagers have access, tragedy is lurking just around the corner.

SECURITY BASIC

SET SOME GROUND RULES There are a few prime directives you have to follow in business: Buy low and sell high, always make payroll, and don't make a mess where you expect to eat. In other words, if you're selling merchandise on Facebook or eBay, use a dedicated computer for those transactions, and use it for nothing else.

Imagine that you are a taxi driver. You wouldn't want to use your taxi to drive down to the corner store and get milk and bread, or take it on a trip across town running errands, would you? The taxi would be your business vehicle, and you would use it only for business. The same goes with your digital storefront. Make it exclusive to the business, and only use it for that. This can limit any potential exposure to bad things, especially since this is the primary place you buy ads, sell products, and interact with money.

HOME-BASED BUSINESS BASICS No matter how small your business, safeguarding it is critical. You might ask, "Who would ever target me?" The answer is "criminals." They can attack you with great ease and with an efficiency, as our forward-looking leaders used to say of nuclear power, that is too cheap to meter. There is actually a search engine on the internet that does nothing but map the machines that are sitting there connected to the internet—it's called Shodan, and it is just one way that your internet-connected refrigerator is in fact known to hackers and vulnerable to attack. Protect yourself.

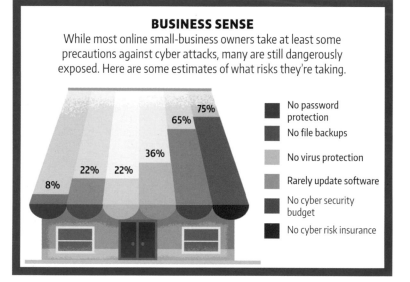

BUSINESS SENSE
While most online small-business owners take at least some precautions against cyber attacks, many are still dangerously exposed. Here are some estimates of what risks they're taking.

- 8%
- 22%
- 22%
- 36%
- 65%
- 75%

- No password protection
- No file backups
- No virus protection
- Rarely update software
- No cyber security budget
- No cyber risk insurance

Build a Wall The first thing you will need is a business-grade firewall. This doesn't need to be expensive—you can get a good-quality home-business firewall for about $200 USD. But business firewalls do some things that home firewalls don't do—or at least do them better: opening ports for virtual private networks, examining packets, and providing services that are above the needs of home users.

Stay Connected You'll need a good, solid way to connect to the internet. Cable modem users might find it best to buy their own (I use an ARRIS modem, made by Motorola), as opposed to leasing one from the cable company, because the connection is actually better and a little faster. Also consider a managed DNS service. This allows content filtering for your employees and can keep malware activity to a minimum. You'll need a decent-grade Wi-Fi access point that supports at least WPA2 pre-shared key (PSK) encryption. Use a strong key and rotate it at least semiannually to keep terminated employees off your network when they are no longer employed.

Let Business Be Business Keeping your work separate from other affairs is about maintaining an atmosphere of professionalism and safety. Do not let your kids or nonbusiness users or traffic on your business network at all. And buy a few hours' help from a local computer outlet—it's worth it to have a checkup on your business network and computer, as well as any necessary maintenance. The silliest thing to come out of the mouths of most smart people is, "I don't have anything worth stealing."

SECURITY BASIC

GROUND RULES FOR SELLING ONLINE Give me your first name, your occupation, and your home city, and I can likely find out almost anything about you in a few searches. Give me your email address, and I can get your Social Security number and full credit file in minutes. And I use legal sites. Imagine what criminals can do.

Even when selling something like a barbecue grill through local ads, be very careful about what information you reveal about yourself. If you're engaging in an online transaction or starting an online dating profile, consider creating a brand-new anonymous email address. Use your first initial only, and be careful about what information you give out. Does Craigslist really need your full address to list a cabinet you're selling? Of course not. Engage in emails with the person you are meeting, and trust your gut about what feels wrong—listen to your instincts, because they are correct.

If you get to the point of a phone call, consider a burner app (available on mobile phones) or a burner phone. Be slow to hand over your number, and agree to meet in public first. Again: Your gut should be in the driver's seat. Make sure someone knows where you are going and with whom you plan to meet—leave a bread-crumb trail in case something bad happens. Consider meeting at a sanctioned online transaction zone set up by local police departments for in-person transactions.

GOOD TO KNOW

DON'T WORK WHERE YOU PLAY When you are working at home, definitely think twice before engaging in any business activities on the same network that your teenagers are also using to play online multiplayer games at the same time. In fact, you should probably just keep your business traffic out of all other networks, and keep all other traffic out of a business-based network. The best way to do this is through a virtual private network (VPN), which is essentially an encrypted tunnel through which your business traffic is shunted back inside your business firewall and then out onto the internet at large. Another option is a mobile Wi-Fi hot spot that only you are able to use and that you will be using for business and only for business. The advice from the movie *Ghostbusters*—"Don't cross the streams"— should suffice here to keep you out of trouble.

GOOD TO KNOW

IMPROVE YOUR INSURANCE You shouldn't assume that your homeowner's or renter's policy will cover you and all of your equipment that's related to your small business—quite often it won't, if you are genuinely running a business from home. And it obviously won't cover your data losses, no matter what. You'll need some type of special business-related insurance rider to cover all the good stuff.

You'll also do well to get your business set up with surveillance cameras that can store video in the cloud (off-site) and that will allow you to view the videos remotely. Should you ever happen to become burglarized (or vandalized or worse), the first question the cops will ask you is, "Do you happen to have any surveillance video?" If you are able to answer their question with a yes, they will become noticeably more interested in their work.

STAY SAFE One exciting way to meet some of the less-scrupulous people in your neighborhood is to install a lot of expensive business equipment in your house. The realities of life in America mean that computer gear provides thieves with tempting targets, but there are steps you should take to protect the most important assets: yourself, your family, your data, and your equipment—in that order.

Don't Advertise You should definitely place your office out of public view, and try to keep the shades drawn and equipment out of sight from outside when you are not around. Invest in an alarm system that has sensors on your windows and doors as well as panic buttons, and the alarm company you choose should monitor the house 24/7/365 and be able to call you or the police at the first sign of trouble. Not only is this great comfort, but it has the benefit of deterring thieves (they see the signs and move to easier prey) and lowering your homeowner's and business insurance costs.

Save the Data Next up is your data. As the veteran of an office that was burglarized by thieves who stole our computers and our backup drive, I heartily recommend cloud-based backup. Don't skimp on a backup solution, especially in an era of ransomware—your backups are what will save your company. Consider turning on Bitlocker for Windows or File Vault for Macs to stop thieves from harvesting data from your stolen goods. This feature for Windows requires the Professional version.

Work with Employees For those who are considering hiring other remote workers, your considerations will be all of the above, plus ensuring you maintain control over shared data. And when you part ways, you'll want to be sure to get your equipment back and "de-provision" them (lock them out of your network and third-party applications).

SORRY BOSS, MY BAD

Employers are put at risk by their staff every day. Here are some of the most common goof-ups that can expose a company to risk.

- Personal use of work phone
- Connecting to public Wi-Fi
- Using the same login for everything
- No cyber security training

63%
94%
49%
45%

SMALL BUSINESSES AND HACKERS Larger businesses and corporations may get a lot more visibility in the press when a hack happens, but that doesn't mean criminals aren't targeting small business every day. In fact, smaller businesses occupy a certain sweet spot for cybercrime. That's because they have more information and assets than a singular consumer, while also being unable to afford as much in terms of security as larger companies.

Just like real-world thieves, cybercriminals will happily take everything they can if they break in, including financial information and records that belong not just to you but to any consumer or client registered with your business—and all of those stolen identities can be used elsewhere. Even your business machines can be locked up with ransomware or infected and drawn, zombie-like, into a botnet for other hacking misdeeds.

So, what can you do to protect your business, its data, and your clients from the effects of a data breach? Quite a lot, actually.

Get Insured Aside from the various security measures we've already covered, cybersecurity insurance is an important part of covering all your bases. Make sure that the policies you consider will cover first-party liability (costs from a breach, legal fees, interruption of business, customer notification, and public relations) and third-party liability (to protect you should your company be involved in a breach that exposes sensitive information about others).

Purge Regularly Never retain business data longer than you must. Once data is not needed, or older than, say, a year, delete it all. Credit card numbers are highly regulated: Understand your obligations under the Payment Card Industry Data Security Standard (PCI-DSS).

Stage a Drill Make it a practice to, well, practice your business's response in the event of a breach. Take the time to formulate, review, and (with each drill) update your response plan as necessary. Perform these exercises at least quarterly, look for any errors or holes in your plan, and then fix them.

Prepare for the Worst Should your business end up the victim of a hack, get to work immediately to find out exactly what happened and put a stop to it, whether that means software patches or a complete takedown and cleanup of your system. Restore any damaged software and documents from backups. Contact your insurer, and get legal advice if you must. Inform your clients as soon as possible of the breach and its nature as it relates to them.

SECURITY BASIC

BACK IT UP Be rigorous about backing up data and storing it in separate places. The best bet is to store things locally, as well as in the cloud. Small businesses can use commercial cloud solutions like Dropbox, SpiderOak, or Backblaze (about $100 USD for two years with unlimited data). Augment this with daily (or more frequent) snapshots of your environment stored elsewhere. A good rule to remember for your business backups is 3-2-1: at least three total copies of your data, two of which are local but on different media (maybe a USB drive or a network-attached storage device) and at least one copy off-site. And be aware the backing up is just your first step. At some point, you will want to test to see if you can restore files. Businesses have thought that they were backing up data only to have some level of corruption ultimately invalidate all the hard work they'd put in.

NEVER CHEAP OUT ON BUILDING AN E-COMMERCE SITE. LOOK AT THE MONEY YOU SPEND ON A PROFESSIONAL AS INSURANCE AGAINST CATASTROPHIC BREACHES.

GUARD A GROWING BUSINESS At last, your business has taken off. So, how do you scale up? The biggest problem for businesses as they grow is that they tend to continue using tools they have instead of reassessing their needs. This is especially true with spending the money to upgrade computers, because business owners don't want to feel like they're getting ripped off: "Why do I need a new firewall? The old one hasn't burned up yet!"

Cyberincident responders see this all the time, even in publicly traded companies—especially ones that have grown quickly. Failures usually fall into two categories. The first one is just that: a failure to properly scope technology requirements and scale technology purchases to match. The second is more insidious: its fast growth, during which executives make decisions to build now and scramble to secure later.

This is the most tempting thing in the world. I can tell you from the painful personal experience of having to break the news to senior executives at Fortune 500 companies that, because of decisions like that ten years previously, the cost to implement a fix is approximately 100 times what it would have been to do it right in the first place. In the security industry, we call this "technical debt," and it's like using a high-interest credit card. You can go ahead and do it, but the day absolutely will come—we guarantee it—when the bank wants its money, and you find out about the wonders of compounding interest. Here's how to avoid that pain.

Get Upgraded Find trustworthy, well-referenced security companies near you and ask them to help you review your needs and make recommendations. These companies are generally easy to check out, and better firms have principals who regularly speak at security conferences, consistently publish articles and write books, and participate in the business community. Find two, and if they generally agree, then go with the one you like most.

Get Tough You'll be looking at a beefed-up version of the secure home office we described. Seek out some kind of centralized authentication system, regular incremental backups and frequent snapshots of your environment, and encryption in every place it can fit. You'll need to hire a good information technology person to manage these systems, or employ a company to manage your infrastructure as a service, an increasingly popular option. You should also consider keeping your critical systems such as servers and firewalls under some form of maintenance contract or support by the vendor. Vulnerabilities in hardware, firmware, and software are constantly disclosed, and having maintenance will keep you up to date without having to either buy new equipment or learn from technical debt bankruptcy.

Get Backup This should go without saying, but back up all your business data . . . and then create a second backup, preferably off-site in a cloud storage server, for example. And then consider a third— just in case. This may sound a little paranoid, but if something goes drastically wrong and you need to restore data, you'll be glad you did.

Get Outside Help Consider hiring a third party to monitor your firewalls and other security gear for signs of trouble. You should have your internal tech person run vulnerability scans regularly so that you keep complete lists of what you actually have and what connects to your network (you won't believe how difficult this is for many companies). Have that double-checked by an outside firm, either regularly as a service or at least once a year.

Get a Checkup Every eighteen months or so, perform a security architecture review. Take this opportunity to reexamine the single most vulnerable part of most business networks: your assumptions. Although taking this step might mean having to pay a bit more up front, it'll be well worth it for the peace of mind it'll bring. As someone who sends his kid to a private boarding school, I can tell you that the alternative is dramatically more expensive in the long run.

KEY CONCEPT

BRICKS AND MORTAR
Any brick-and-mortar shop that has developed a great online presence gets a great benefit from the fact that the cost of maintaining this infrastructure is much cheaper than it used to be: Now, you park the entirety of your business's electronic infrastructure in the cloud instead of maintaining it on-site. But "cloud" does not mean "secure." Cloud infrastructure still has many of the issues of traditional infrastructure, and if you're selling goods online, you have issues of payment card industry compliance to deal with as well. Fortunately, most of this can be outsourced, reducing costs even further. However, it pays to have third-party firms provide you with vulnerability analyses, architecture reviews, and penetration tests—especially against your primary business applications—regularly.

BACK UP PROPERLY

If utilizing a Windows server, you can turn on shadow copies, which allows you to revert file changes made to the system. This can be done by starting the volume shadow copy service and changing the startup type to automatically via the services in the administrator control panel in the control panel. You can then go to File Explorer, right-click on a drive, go to the shadow copies tab, and enable the option on the drive or on other drives on the system. Shadow copies aren't backups but a quick way to revert files or folders, such as when a user accidently deletes a file or if a few files become encrypted by a crypto malware. You'll need professional help in setting this up, but on the day ransomware encrypts all your files, or all your machines get stolen, it will be worth every penny you spent and then some.

ONLINE BUSINESS BANKING Business banking is where most companies get into trouble. The problem is that banks are not actually responsible for wire transfers and automated clearinghouse transfers made from your business account without your knowledge provided that those transfers were made using your credentials. Business accounts are not protected in the same way that personal ones are. Most banks will still try and help you claw back stolen money, but in some cases where the victim of fraudulent transfers sued the bank, the bank turned around and counter-sued that customer . . . and won.

Keep It Simple Unless you employ more than, say, 250 people, we recommend not using online business banking without an actual two-factor authentication login and a voice verification from the bank. Weirdly, the only banks that seem to offer this without any hassle are not the big guys but the boutique banks.

Get Personal We have found that these smaller banks—the ones that provide personal bankers who know you and your voice, and who ask about your family—are usually the banks with the best possible security you can get. The argument about availability of bank branches is absolutely valid—but with the security upside of the most common vector of attack (wire and ACH fraud) handled, the inconvenience may be worth it. Oh, and we have found that the costs are just about the same between the big banks and the boutiques.

No matter what kind of online banking you prefer, we recommend using a dedicated computer for it. The less that computer interacts with the public internet, the less of a chance there is of your credentials being hijacked.

A SECURE WEBSITE A business without a website is simply not taken seriously. Where people get into trouble is trying to build an e-commerce website on the cheap. That is a guaranteed disaster—note, I didn't say almost a guarantee. It's rock solid. If you want to sell things on your website, you need to understand that one of the internet's most vulnerable things is the commercial web application.

The good news is that if you're just looking for a basic web presence—a site consisting of a home page, contacts, information about your services or products, and that sort of thing—there are a lot of very cheap options out there that are in fact quite beautiful and professional looking. Right now, I'm partial to Squarespace. They have designs you can personalize, and you pay a monthly charge for them to host it. They even make it very easy to buy your domain name—all these things that vexed your predecessors in the 1990s and aughts have been reduced to wizard-based menus.

Beyond any design aesthetics and good customer service you've built in, there are also some important guidelines to creating and maintaining a properly secure business website.

Stay Secure First, make sure your site itself, when built, is a secured one. This means your site's address will have an https prefix instead of just http, along with a small padlock icon. This adds a layer of encryption and makes it harder for hackers to break in. Further layers of security, such as a web application firewall (WAF), will add to your protection. A "secure and verified" badge added to your site, when clicked on, will also provide full verification to visitors, including the date of the last security scan.

Keep It Up to Date Whether you build your site yourself or trust someone else to do it, keep your software updated. New exploits are found on a weekly basis, and you don't want to become the latest victim of something like the 2017 WannaCry attack in which computers that hadn't installed a basic security update were hijacked to attack hospitals, phone companies, and others in 150 nations.

Run a Tight Ship Change passwords often, and keep them strong. Hide and rename admin directories in your business website to thwart any hackers, as they invariably will go after files and folders with names such as "admin" and "login."

Be Transparent Display your business's privacy policies on your website, explaining what data is collected, how it is secured, and what is done in the event of a breach; update your policy as needed.

GOOD TO KNOW

A STRONG DEFENSE IT security is best done in layers; the more there are, the harder it is to have unauthorized access. No matter the OS, no single silver bullet will keep you secure. So, what to do? Turn on the firewall built into the OS. Inaccessible application port connections reduce security risk. Update software regularly, not just automatic updates, and keep antivirus software to protect your machine and data, including against malware. Disable remote access if unneeded. Use 2FA and VPNs through a company firewall with support agreements (to add content filtering, remote access, constant updates, and malware protection). Keep Wi-Fi separate and protected, and employ intrusion monitoring along with virtual LANs. Control with whom files are shared, consider security awareness training, and, last, keep all hardware in a secure room with restricted access.

GOOD TO KNOW

REPUTATION REPAIR

Let's say people are leaving bad reviews of your business on Yelp or Google. Instead of hiring services to try and "fix" your reputation, go online and answer questions and concerns. If a negative review on Yelp is followed by your side of the story and an offer to make things right, people tend to cut you some slack.

Real, useful content that shows your business in a positive light is often weighted more than complaints and slander by search engines. That said, if you're an executive who said something untoward on Twitter, for example, and the story is picked up by the press, you're in for a few years of reputation building. In this case, a reputation repair service might take some of the weight off of you. But there is no way to erase stuff from the internet: Work on new, relevant, accessible content that search engines will value more than the bad stuff.

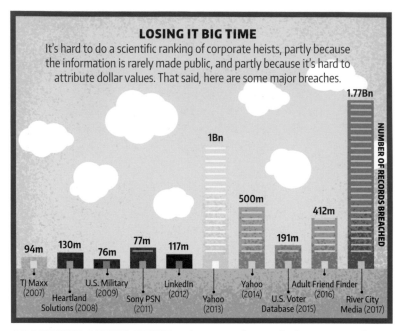

LOSING IT BIG TIME
It's hard to do a scientific ranking of corporate heists, partly because the information is rarely made public, and partly because it's hard to attribute dollar values. That said, here are some major breaches.

NUMBER OF RECORDS BREACHED

- TJ Maxx (2007): 94m
- Heartland Solutions (2008): 130m
- U.S. Military (2009): 76m
- Sony PSN (2011): 77m
- LinkedIn (2012): 117m
- Yahoo (2013): 1Bn
- Yahoo (2014): 500m
- U.S. Voter Database (2015): 191m
- Adult Friend Finder (2016): 412m
- River City Media (2017): 1.77Bn

ORGANIZED RETAIL CRIME When people watch the news and see a breaking story announcing that, say, fifty million peoples' credit card data was stolen from Target or Home Depot, they sometimes wonder, "Just what the hell can you do with fifty million stolen credit card numbers?" The answer is simple: You can sell them to other criminals—in blocks of 1,000 card numbers at a time.

Criminal Coding Criminals can monetize stolen cards in any number of ways, but one popular method involves gangs of crooks who buy a bunch of stolen credit cards and a magnetic card encoder (similar to the kind used to make hotel room keys). They pay people to obtain cards of all kinds with magnetic stripes: used gift cards, used hotel room keys, stolen credit cards—anything with a stripe. Then the perps re-encode them with the information from the stolen cards.

A Larcenous Road Trip Once the cards are coded the gang rents a moving truck, assembles a small team of accomplices, and drives along an interstate highway until they come to a shopping mall. After giving each of the team a little stack of the re-encoded cards, the leader sends them on their merry way into the mall, with instructions to use those cards to buy things under a certain amount—say, $500 USD—and bring them back to the truck. If a given card doesn't work, they simply throw it away and swipe another one, working their way through the mall and through the stack of cards at the same time.

When they're done at one mall, they drive to another. Soon, they have a truck full of swag. And at some point, the swag gets sold on eBay or other online outlets.

There's big money to be made, and that's one of the reasons banks and credit-card companies rolled out chip + PIN cards, designed to combat exactly this kind of point-of-sale scam. This development means that criminals will have to find some new way to rip of card-holders and businesses. I have faith in their ingenuity.

THE TAKEAWAY

Head spinning with all the ways someone might try to rip off or otherwise undermine your small (or not so small) business?. Check these basics.

BASIC SECURITY	• Encrypt, back up, and use strong passwords for data. Hire professionals to help set up systems. • Delete as much information as you can every week; you can't lose data you don't have. • Use a cloud-based service to back up your minimal set of business data. • Teach your employees best security practices and policies. • Train any employee authorized to transfer money about inviolate procedures for wire transfers.
ADVANCED MEASURES	• Employ two-factor authentication and do vulnerability and penetration tests regularly. • Take snapshots of all computers at least daily, if not more often, and store them encrypted in the cloud. • Have a plan in place in case any of your business-related devices are compromised, lost, or destroyed. • Run quarterly tabletop exercises to practice the plan and find problems with it.
TINFOIL-HAT BRIGADE	• Keep single-sign-on, forced-VPN, all-virtual desktops. • Run phishing awareness and other security programs as training for your staff. • Keep an incident-response company on retainer.

HALF OF ALL SMALL BUSINESSES HAVE A SECURITY BREACH EACH YEAR. PROTECT DATA WITH SIMPLE ENCRYPTION AND GOOD PASSWORDS. A BREACH RUINS CUSTOMER TRUST. SPEND BEFORE, NOT AFTER, THE BREACH.

THE FUTURE OF MONEY

THESE DAYS THE PHRASE "NEW MONEY" HAS COME TO MEAN A PORTFOLIO OF VIRTUAL CURRENCIES. IN A WORLD WHERE MONEY IS JUST A SERIES OF ONES AND ZEROES, HOW DO YOU KEEP YOUR ELECTRONIC PENNIES SAFE?

Throughout history humankind has relied on some form of currency, using everything from shells to gold to make purchases. The island of Manhattan was even bought with a bag of beads not so long ago. We use it every day, but we tend not to think much about the concept of money beyond whether we have enough. In fact, the idea that we've agreed that these pieces of paper and metal represent real value that can be traded for pizzas and underpants is quite remarkable. Over the centuries, things have changed at a blinding pace in the world of money, and it won't slow down anytime soon.

In the 1960s and '70s, modern credit cards were introduced, bringing with them a whole host of new and exciting ways for criminals to steal your money. Some banks were so trusting of their customers that they didn't even do credit checks . . . they just mailed out these shiny new cards and assumed that the recipients would be sensible. That went about as well as you might imagine. And that was before anyone had the technology to skim those digits, recode them onto a hotel card key, and go to town. Now, digital banking services like PayPal and ApplePay, and the brave new world of digital currencies like bitcoin (BTC), mean there are even more ways to steal your scratch.

CASH FREE I recently went to Australia on a two-week business trip. As an experiment, I wanted to see if I could go the entire trip without using any Australian currency. To my surprise, I discovered that I could! I used my credit card for the majority of my transactions. (To Australia's credit, it has one of the most innovative payment industries in the world; the country has adopted new payment technology that makes a purchase as easy as waving your credit card over a terminal.) I used various apps on my mobile phone to pay for my transportation. Each of these apps had previously saved my payment details, so they automatically billed me in my native currency. At one point, I actually needed cash for a purchase, but I just happened to have some U.S. dollars in my wallet, and the small-business merchant agreed to accept them.
—Heather Vescent

LIVING ON BITCOINS In 2013, reporter Kashmir Hill wondered whether she could live for a week paying for things only in bitcoin. She had trouble finding vendors who would accept Bitcoin for services and many hadn't heard of the new currency. When she tried the experiment the following year, after several start-ups had made it easier for merchants to accept bitcoin, she was able to use a ride-sharing service and pay for a seventeen-course dinner. So yes, while you can purchase cupcakes, lunch, coffee, and even a mani-pedi on Wilshire Boulevard in Los Angeles with bitcoin, it's not the norm by any means. It's still mostly a novelty for many, although there was plenty of buzz about bitcoin a few years ago, when it was at the peak of the hype cycle. And today there are those who believe that digital currency can still change the world.

A CURRENCY REVOLUTION The idea of cryptocurrencies kicked off in 1998, with an essay written by computer engineer and cypherpunk Wei Dai that envisioned a form of money created and controlled by cryptography instead of a central government. In 2008, a pseudonymous developer operating under the name Satoshi Nakamoto released an open-source white paper describing a peer-to-peer method for creating a cryptocurrency called bitcoin. In the beginning, the only attention it received was from a small group of crypto enthusiasts. Now, bitcoin has kicked off a currency revolution and re-envisioned money for the information age.

How Is Bitcoin Made? Unlike physical money, no one person or government owns the technology or concept behind bitcoin. New bitcoins are generated by people on the internet who competitively work to record and verify previous bitcoin transactions—starting from the very first transaction in 2009—in a process called mining. Think of it as using your computer's processing power to solve a complex puzzle. Whoever solves a step (or block) in the mining process is rewarded with a new amount of bitcoin.

Is Bitcoin Actually Money? The U.S. government sees bitcoin as a commodity rather than a currency, but many use it for transactions as if it were money. Its value fluctuates, generally trending up as the number of new coins decreases, and acts like a stock at times, rising and falling based on perceived value. In 2009, the first bitcoin was valued at $0.07 USD, but it soared as high as $1,250 in 2017.

Will Bitcoin Replace Other Currency? Early on, many speculated that bitcoin could replace nation-state-backed money such as the U.S. dollar. Once it was released into the wild internet, it took on a life of its own and attracted a different set of users, forcing governments to react (with the U.S. government classifying it as a commodity, for example). Nevertheless, it started a revolution—showing a viable new way of creating currency for the digital age.

Is Bitcoin Anonymous? One of the "features" of bitcoin is the ability to complete anonymous transactions. But this isn't totally accurate. Financial institutions require compliance with KYC ("know your customer") regulation (see page 123), so if you buy bitcoins from an exchange where you have a connected financial institution, your wallet can be traced to your identity. However, there are ways to set up anonymous wallets. In either case, the transactions are recorded on the blockchain (see page 120).

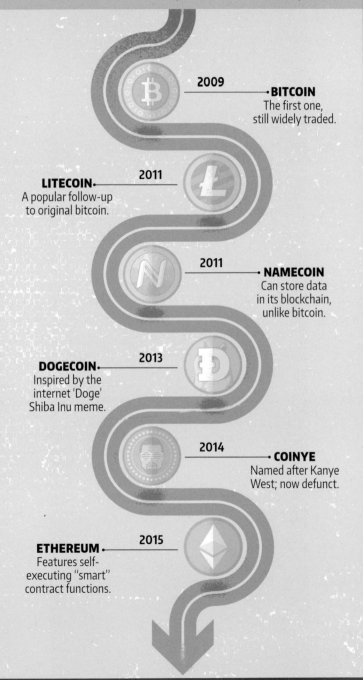

NOTABLE CRYPTOCURRENCIES

There are hundreds of cryptocurrencies out there. Most are copycats or have been developed to expand initial bitcoin functionality. Here are a few examples.

2009 **BITCOIN**
The first one, still widely traded.

LITECOIN **2011**
A popular follow-up to original bitcoin.

2011 **NAMECOIN**
Can store data in its blockchain, unlike bitcoin.

DOGECOIN **2013**
Inspired by the internet 'Doge' Shiba Inu meme.

2014 **COINYE**
Named after Kanye West; now defunct.

ETHEREUM **2015**
Features self-executing "smart" contract functions.

TRUE STORY

ELECTRONIC HEISTS

New currencies, like all new technology, are experiments and not secure or fully tested. When you put your skin in this game, you could lose it. In 2013, Bloomberg TV hosted a bitcoin special, wherein host Matt Miller gave anchors Adam Johnson and Trish Regan a code that would give them $20 each in BTC. A fast-thinking viewer used the code on the TV screen, effectively snatching it from Johnson's hands when he showed it to the studio camera.

Larger losses have also happened. Mt. Gox, based in Tokyo, was one of the first bitcoin exchanges. At one point, it handled 70 percent of the world's bitcoin trades. In 2013, it took up to months for customers to withdraw funds, halting operations several times before closing in February 2014. Through bankruptcy filings, it was discovered the company stole as many as 650,000 bitcoins from customers.

KEY CONCEPT

THE BLOCKCHAIN All bitcoin transactions that take place are recorded on the blockchain—a database that acts as a public ledger and helps to reinforce the cryptography behind the currency. Transactions are each recorded with a time stamp, the amount of transaction, the wallet address that sent the bitcoins, and the address that received it.

With cash transactions, no one knows the details of your actions. With bitcoin, a certain number of transactions are formalized into a "block." When each new block is recorded on the previous block, the transaction data is set in stone and can't be changed. The blockchain is the ultimate tracking system—it's decentralized, no one can manipulate it, and any information added to a blockchain is also permanently recorded. Because the blockchain records are set in stone, it reduces the bitcoin's full anonymity.

MONETARY MYSTERIES AND ELECTRONIC INTRIGUE As cryptocurrencies continue to develop, so too do the questions behind them. For example, no one has yet discovered the true identity of Satoshi Nakamoto—possibly a pseudonym for a single person or a collective group who worked on the idea of bitcoin together. Whether Nakamoto is an individual or a group, the introduction of bitcoin has changed the way we look at money and its relation to the internet forever.

Investing in Cryptocurrencies Those who are curious about bitcoin and other cryptocurrencies can acquire them in multiple ways. The original method is to join in the mining effort, which means using your own computer and a set of specialized software (and sometimes extra hardware) to work on the blockchain, and thus unlock the next set of bitcoins by solving it. Don't be surprised if it takes a while, though: Mining on your own is akin to playing the lottery, while mining in a group (or pool) means getting a return equal to the fraction of the pool's computing power that you've put in. Multiple exchanges are online, and you can also purchase cryptocurrencies from other people.

No matter the method, you'll have to use an address—a public string of numbers—to send or receive bitcoins, similar to the way that an email address handles messages. A wallet is actually just a private string of code that corresponds to the address, and stores the cryptocurrency info, keeping the bitcoins safe and reached only to the person who has access to the wallet. The wallet is not usually

IMAGINARY MONEY

For bitcoins or other cryptocurrencies to work in the market, they need a level of stability and buy-in. Here's how that works.

Bitcoins go from an anonymous wallet... ...to a tumbling site... ...Broken into multiple transactions... ...Amounts confirmed in the blockchain... ...Funds are then recombined... ...And deposited in another wallet.

physical—although some people do indeed keep access to these digital wallets on a physical object, such as a USB stick, to reduce the risk of losing cryptocurrency in a hack.

Convenience or Anonymity? If you buy bitcoins from an online exchange, they will give you your own address and wallet. Once you have connected your bank account and once it's been confirmed as yours, you can buy bitcoins from the exchange, storing them there on their in-house wallet, or you can export them to an outside address. As an investment, bitcoin and all cryptocurrencies are high-risk. Bitcoins have been stolen, and legitimate exchanges have gone bankrupt and customers have lost their bitcoins. In order to have a totally anonymous bitcoin wallet, you will have to resort to buying the bitcoins in person—yes, this means that you'll have to physically hand someone cash. They will then send bitcoins to your anonymous wallet. Since cash transactions are not tracked, you can have them transferred to a wallet that has no identity associated with it. (The electronic transaction will be recorded in the blockchain, regardless.)

Bitcoin and Black Markets Up to this point we have discussed the legitimate uses for bitcoin. But plenty of people out there also utilize bitcoin for illicit transactions, money laundering, or moving money around in ways that can't be easily tracked. There are innumerable black markets, and the most popular and active ones are always changing as old ones are shut down and new ones pop up. The first and most notorious online black market was called Silk Road, and it was started by the Dread Pirate Roberts in 2011. Functioning much like eBay, Silk Road offered illegal and prescription drugs, hacked data, fake IDs, and more. Silk Road was shut down by the FBI in late 2013, and the Dread Pirate was unmasked as Ross William Ulbricht, who is now serving a lifetime prison sentence (see pages 174–175 for more on his strange tale).

The Future of Cryptocurrency In all likelihood, bitcoin and other cryptocurrencies will stick around for some time to come. The way has been paved for modern digital currency experiments, putting pressure on the traditional financial transaction methods to reduce bank and transfer fees, while increasing transfer speeds. Although we will see more characteristics of this technology as part of our existing currencies in the future, it won't entirely replace the U.S. dollar for groceries and gas anytime soon. Cryptocurrency values fluctuate too frequently, governments see them as commodities at best, and the technology will continue to be targeted for hacks.

T/F

YOU CAN LAUNDER MONEY WITH BITCOIN

TRUE When you use cash to buy bitcoins from someone in person, there is no trail. Bitcoin wallets are not required to have identifiable information. An anonymous email address can be used to start a wallet, which can then hold bitcoins. If you end up with a large sum of bitcoins and want to cash out, you can find someone to exchange them. The trick is to keep identity data away from these bitcoin wallets. Cash transactions aren't recorded, so it makes using bitcoins to launder money appealing. There are also "tumblers," programs or sites that mingle fractions of BTC in multiple transactions. After a time, the bitcoins come out clean—well, at least in theory.

MOBILE MINUTES AS MONEY Currency is essentially a physical representation of value. Historically, gold and other precious metals and minerals have been used as a medium of exchange based on their rarity and perceived actual value. It makes sense that anything of mutual value between parties can be used as money. In several parts of Africa, technology has taken the place of bills and coins: Savvy entrepreneurs have been using mobile airtime as a currency, such as the M-Pesa system in Kenya, which allows the exchange of money and airtime. Prepaid mobile phone cards are used instead of bills, and mobile phone owners rent out phones or barter minutes for goods and services. Given their near-universal utility, mobile minutes are considered more sound than many regional currencies subject to a government's stability or economic fluctuations.

TECHNOLOGY AND MONEY As our technology evolves, so will our finances. From IOUs to ledgers to consistent coins and bank notes, it makes perfect sense that as our technology improves, we would apply it to money.

Smartphones and Finances Anyone with a smartphone basically carries a tiny computer in their pocket and can connect to the global financial network. Whether you are making a bank transfer, sending a PayPal payment, making a trade, buying a tomato at a farmers' market, calling for a ride share, or paying with bitcoin through a QR code, today's mobile phones are more sophisticated than the bank tellers of twenty years ago.

Credit Cards One of the money dreams of the future is the creation of a single global unified currency. We already have that today in a way, thanks to credit cards that are accepted almost everywhere and have excellent fraud protection. But even with the best security system, sophisticated social hackers can successfully impersonate you despite the strictest precautions. Credit cards are never going away and are a de facto universal currency.

Blockchain Expansion While they may sometimes vary in perceived value between curiosity and commodity, cryptocurrencies aren't going anywhere anytime soon—and neither is the blockchain. In fact, the concept of the blockchain can be applied to other parts of the world's economy, and not just for tracking of cryptocurrencies and the transactions they are used in. As the transactions on a blockchain are set in stone once recorded, the technology could be used to create cheap, tamper-proof public registries of who owns various land or property, notarize documents without the need for a notary on-site, and even ensure the security and value of stock market and other high-value trading systems and financial transactions.

Smart Contracts Ethereum, one of the more notable cryptocurrencies, was created by Vitalik Buterin, who wanted to build applications on top of the (bitcoin) blockchain. Since he could not add this extension to bitcoin, he launched Ethereum. The big innovation with this cryptocurrency is its ability to write contracts—bits of code execute on their own when certain conditions are met. On Ethereum, developers can write their own smart contracts. In the future, we can imagine a world where a piece of software initiates a financial transaction once certain conditions are met. Smart contracts could be used for human transactions, but more likely they will be used for computer-to-computer payment transactions.

Financial Intelligences It's not far-fetched to imagine software making transactions on your behalf. Automated bill paying already exists, which lets you set your bank account to make payments on your behalf. As we develop better intelligence, personal financial intelligence may someday go further.

In the future, we can expect our technology to take care of the financial transactions entirely. This is something that is already happening with services like Airbnb, Lyft, and Uber, which have the payment aspect built into the software app experience, removing it completely from the real-world experience. Amazon has done something similar with its innovative physical stores, where you can walk out with the items you want and are automagically charged for the items you selected instead of having to check out.

In the end, the future of money must include more security and convenience while being easy to use. As technology continues to innovate, we will also have to keep up with bugs, flaws, and loopholes, fixing them as they arise.

KEY CONCEPT

KNOW YOUR CUSTOMER Any financial institution out there has to have a degree of security, stability, and trustworthiness in order to operate, and that includes having stable customers. Traditional banking regulations require banks to know their customers (called "KYC" in bankspeak). To comply with these financing regulations, banks have to confirm the identities of all their account holders. When you open some bitcoin wallets, especially any that are connected to traditional bank accounts, you may have to prove your identity. It's not recommended to do illicit transactions or make black-market purchases with these accounts—unless you want to increase the chance of being busted. However, there are other ways to acquire bitcoin that do keep the purchases anonymous and can also facilitate money laundering.

SMART MONEY, SMART THIEVES
You might think that increasing surveillance and face-recognition software would help catch the guy who stole your wallet, especially with all the cameras out there. But the problem is accessing the data before it's deleted and then taking action on it.

In the case of Quentin Hardy, whose wallet was stolen in San Francisco, the thief used his stolen credit card for Uber. The card was connected to an existing account, and since Uber keeps data on all its rides, it had info about the thief—including GPS data that might show where the thief lived. The problem with this is getting the data, which often can only be legally released to the enforcement, and the infraction has to be big enough to warrant an investigation. Often, it is not. And even so, by the time the police get around to seeing the video footage, it may have already been deleted.

DIGITAL SAFEKEEPING In the future, we won't get to simply stash our cash in a safe or under a mattress. New forms of finance mean new protocols for keeping your money safe.

Keeping Online Integrity While many governments offer individual reimbursements from fraud, business accounts are not always guaranteed the same security, and online banking and fraud protection are not typically in the hands of the user. Look for financial institutions that have good security—sometimes it's hard to find out which ones have been hacked, because no one wants to disclose that information. Use robust passwords in online banking, change them often, and don't reuse old passwords. Limit who you share banking authentication credentials with to reduce the chance of unauthorized transfers or transactions.

Stay Secure with a Selfie Banks are motivated to use secure systems. Passwords can be difficult to enter on a mobile device, so banks have other secure authentication systems options: your fingerprint, a PIN, or facial (or even voice) recognition. The secure selfie is even hacker proof—you must blink or make a facial gesture that you can't duplicate with a photograph.

Be Safer with Biometrics Biometric verification has been thought of as the great fail-safe. The idea is that it's near impossible to replicate someone's fingerprint or iris or retina, although hackers (and Hollywood) have shown ways to duplicate a fingerprint. Unlike a password reset, it's not that easy to get a new fingertip or eye. There are, however, security combinations that use hashed biometric data for three-factor (adding a biometric to the password and token) or even four-factor identity verification involving a second biometric. The more factors that are required for a transaction, the harder it is for a thief to steal or replicate them.

Practice Credit Control Use multiple credit accounts. The first is for automatic monthly payments; the second (with a low limit) for spontaneous online purchases and the like (this number can be virtual, or even single use, so you have to get new ones each time you want to use the account); and a third for physical shopping, with a slightly higher credit limit. Monitor your credit score. You'll know if there is a change in your score and be empowered to fix it. Check your bank accounts on a regular basis via a dashboard or phone app, and set up notification for certain transactions and amounts. Tell your bank or credit company when you are traveling so it will allow your purchases. Know your bank's customer service numbers and how to stop payments or authorize a charge on your mobile app.

BIOMETRIC AUTHENTICATION METHODS
You can use more than just fingerprints for confirming someone's identity—and more means are developing all the time.

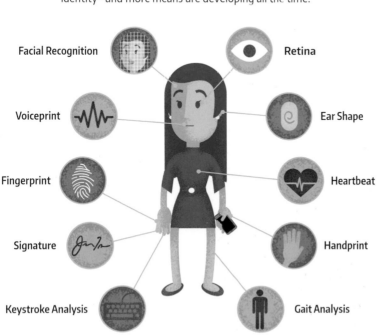

Facial Recognition

Retina

Voiceprint

Ear Shape

Fingerprint

Heartbeat

Signature

Handprint

Keystroke Analysis

Gait Analysis

FUN FACTS

BIG LOSERS Big money plus gambling can mean big losses. You can try to blame it on the free drinks—which is what Mark Johnston tried to do when he lost $500,000 USD one weekend in Las Vegas, saying he was black-out drunk—to avoid paying gambling losses. Then there's the story of the CEO of the Oriental Trading Company, Terrance Watanabe, who lost $127 million USD at Harrah's in 2007 and accused the casino of giving him pain medication and keeping his glass full. (He was also known to play three hands of blackjack at the same time.) Meanwhile, a software bug on a Eurobet website caused Bruno Venturi to make £650,000 from £17—but the company refused to pay, citing the software bug. And Harry Kakavas from Australia lost $1.5 billion AUD in just over a year and attempted to sue, claiming casinos took advantage of his status as a compulsive gambler. Play responsibly!

GAMBLING AND THE INTERNET We would be remiss in discussing the internet and money if we didn't also include a throw of the electronic dice. It used to be that if you wanted to try to get rich quick (or more likely just waste your money), you had to go hit the slots and tables in places like Las Vegas or Atlantic City or Reno—or maybe find a smoky back room for an underground poker game. These days, you only have to go as far as your home computer. That dicey place where the internet and your finances intersect is in online gambling, and while you won't have to worry about ending the night smelling of cigar smoke out of your hair or avoid having a cocktail spilled on your suit, internet poker or roulette or slots carry plenty of their own risks and concerns.

Living with the Law Depending on where you are in the world, local laws restrict gambling online as well as in person, especially in the United States. the Federal Wire Act of 1961 has been construed by courts to mean that it covers all forms of gambling. The FBI may not break down your door if you bet a dollar on that online poker forum, but the legal risks are still present, as well as dealing with the tax issues should you win big.

Supporting Your Local Mob There are plenty of historical ties among gambling, casinos, and organized crime, and internet gambling is no

exception. Plenty of gambling sites are relatively legitimate, but it's just as likely that a given online casino could also be a place where money is being laundered and other shady dealings are being supported behind the scenes.

Going for Broke The biggest risk associated with gambling is losing all your money or developing an addiction, or both. Humans tend to respond strongly to game-of-chance situations, due to a psychological effect called intermittent reward: If you don't know when to expect the payout from a gamble, there's a chance you'll keep trying over and over. And if you can go for that potential reward without having to leave your house and visit a casino, you could easily empty your savings account while still in your pajamas.

Making Your Loss Their Profit Whether the risk is tied to organized crime or no, internet gambling requires spending money, either by transferring funds from a bank account or providing a credit card number or some other means of sharing your financial information. If a gambling site you're visiting gets hacked and you have any information saved there (or any money left in an account on-site), then you run the risk of getting your finances stolen.

THE TAKEAWAY

Putting your money where your modem is opens you up to all kinds of new and exciting financial risks. Here are some methods of staying safer while spending online.

BASIC SECURITY	• Monitor your accounts. • Set up alerts for your accounts and for purchases over a certain amount. • Use a strong password, and don't use the same password for multiple accounts. • Use a stronger password for your financial accounts than anywhere else.
ADVANCED MEASURES	• Use multifactor authentication. • Use credit cards that offer fraud protection and identity protection. • Monitor your credit score.
TINFOIL-HAT BRIGADE	• Use financial services that offer biometrics, or three- (or four-) factor authentication. • "Launder" your cryptocurrency by using services that obscure the source. • Use cryptocurrency like a one-time pad: buy the coin, launder it, make your purchase, and erase all records.

STATE AND LOCAL LAWS CAN AND DO OUTLAW VARIOUS FORMS OF ONLINE GAMBLING. CHECK YOUR LOCAL LAWS, BECAUSE SOME PROHIBIT ANY TRANSACTION INVOLVING GAMBLING SITES.

DEGREES OF DECEPTION

EACH YEAR, PHONY SCHOOLS RAKE IN BILLIONS OF DOLLARS FROM UNSUSPECTING OR EVEN CRIMINAL STUDENTS WHO MIGHT THEN END UP IN CHARGE OF NUCLEAR REACTORS, NATIONAL SECURITY, OR YOUR EMERGENCY HEART SURGERY.

The very first universities that granted academic degrees to their students were established in Europe in the eleventh century. It should come as no great surprise that, not too long thereafter, the first counterfeit degrees began appearing as well. In America, fake schools were awarding degrees long before the Revolutionary War. But why should we fret if a Renaissance count or a gentleman farmer bought the right to call himself "Doctor" without doing any work? Historians are probably the only ones who care about those cases, but everyone should be concerned when, as happened recently, a highly placed (and unqualified) Homeland Security official and a top nuclear engineer turned out to have purchased their degrees. Fake college degrees are held by tens of thousands of deceptive doctors, lawyers, therapists, teachers, and others whom we count on to be well trained. And that doesn't include the thousands more noncriminal but naive degree holders who simply fell for a fast-talking salesperson who convinced them to lay down heaps of money for nothing more than a worthless piece of paper—one that may, in fact, be a lot worse than worthless. A fake degree can be a career-ender if it should be exposed—and that's to say nothing of the risk to others before the truth comes out.

FOREIGN EXCHANGE How do any of these degree mills actually receive legitimacy or accreditation? In one example, in 2003, the government of Liberia—at the time considered corrupt, unstable, and nearly bankrupt—wrote to a number of questionable schools offering them the full accreditation of the Liberian Ministry of Education for the low, low price of $1,000 USD (later raised to $10,000, and then increased once more to $50,000 down and $20,000 a year). In short order, more than a dozen non-wonderful schools, most of them located within the borders of the United States, were proudly advertising their certificates of accreditation—signed by a legit minister of education from a nation that's unquestionably a member of the United Nations. That situation was eventually resolved legally through UNESCO, but there's nothing to stop it from happening again elsewhere.

BUYING FROM DIPLOMA MILLS People purchase degrees for any number of reasons, but the following are the most common.

Employment Opportunities Many jobs will require that people in certain positions have a specific degree (or, sometimes, any degree), which can lead job hunters to fake it. Others believe that the esteem of an advanced degree will give them an advantage in the market.

Professional Prestige While a marriage counselor doesn't need to have a doctorate, a financial planner an MBA, or a contractor a degree in engineering, being able to advertise those degrees may attract more customers or inspire more confidence from clients.

Professional Validation You've worked hard all your life. Why shouldn't you get the recognition you deserve? This is a valid argument, as the practice of credit for "life experience" illustrates. But all too often, this line of thought is used by fake schools to convince potential students that rewarding their hard work with a phony degree is totally logical—on par with earning a real degree via a legitimate proper education.

DUE DILIGENCE So, what if you suspect that someone you've encountered has faked or fraudulently obtained their college credentials? If it's your old Uncle Morty talking up his suspicious-sounding degree in film studies to win an argument, maybe you can just roll your eyes and let it go. But if the phony degree could have real consequences for the health or safety of others, you should check it out. Looking into the details of a school can also give you reason to steer clear of it if it proves to be an imposter.

FALSE WITNESS One of the more frightening subsets of prestige, or being seen as an authority in a particular field, is the matter of expert witnesses in court cases. A lot of damage can be done—but the courts rarely seem to check the legitimacy of those witnesses. Indeed, this author, Dr. John Bear, has provided expert testimony in dozens of cases of academic fraud; to the best of his knowledge, no one has ever checked his credentials. But this negligence has had serious real-world effects.

The Case of the Fake Engineer One particular self-styled automotive engineer testified on the behalf of an auto manufacturer that the brakes on a vehicle involved in a fatal accident "could not have failed." This man was later exposed on the witness stand as having bought his credentials from a notorious degree mill.

Adulterated Architecture 115 people died when a skyscraper collapsed in a 2011 New Zealand earthquake; the building's designer faked his engineering credentials.

Environmental Hazard We learned of an asbestos removal "expert" whose degrees in environmental science came from a worthless source, rendering all of the "improvements" he'd recommended suspect.

Burn Notice A certain burn expert always concluded that the terribly injured burn victims involved in cases where he was asked to testify had caused their own injuries, and that his corporate clients were not responsible. His fake PhD in safety was at last exposed, triggering a review of cases in which his testimony had been crucial.

Liar, Liar One high-profile lie-detector expert who had appeared on *Larry King Live*, *Geraldo*, and other talk shows, and who was also involved with the O.J. Simpson and JonBenét Ramsey cases, held a doctorate from a notorious degree mill.

MY DOCTOR MIGHT BE A PHONY

TRUE Terrifying as the concept is, thousands of people have purchased fake medical degrees and are using them to prey on unsuspecting patients. one of the most horrific examples is that of Michail Sorodsky, who bought a phony medical degree, set himself up as a "holistic healer," and, in 2011, was convicted of sexually assaulting women who came to him for treatment. Some of them later died of untreated cancer. Many other deaths and injuries at the hands of fake doctors have been documented over the years—and those are just the ones who got caught. After all, you've probably seen all those framed degrees on your doctor's wall. But have you ever fact-checked them?

- - - - - - - - -

NONSTANDARD SCHOOLS Not every nontraditional school out there is a bad one—some are just, well, not traditional. One thing to be aware of is that shady schools may well use the cover of one of these legitimate alternatives to disguise their own lack of a proper curriculum.

New Ideas Educational models change—not long ago the idea of earning a degree online was crazy. Now it's standard. But beware of schools that claim that they are too innovative to be accredited.

Religious Studies Many totally legit religious schools consider secular academic accreditation to be irrelevant to their mission.

Geographic Variations Standards and practices vary by nation, and even by state in the United States. Some schools game this by claiming a location most favorable to their business model.

CHECK OUT A SCHOOL Some fake schools are highly sophisticated in their advertising and employ expert salespeople to convince wary would-be enrollees to take the plunge. It's not uncommon, when a fake school is exposed, for its customers to be chagrined, saying they should have known better but the salesperson was so convincing, the catalog so shiny, the website so nicely done that they got sucked in. Of course, one of the best possible ways to check out a school is to make an in-person visit, but if that were cheap or easy, online learning wouldn't be so popular.

CONSULT THE PROPER AUTHORITY In the United States, every state has a higher-education agency or the equivalent; most countries have either a single national ministry or a similar situation. An agency will rarely comment on the quality of a school, but it will let you know whether it is properly licensed. Some fake schools go as far as setting up a fake accrediting agency that, miraculously, only have one client, or worse, also accredits Harvard and Yale (without their knowledge), so make sure you find the relevant office through the official U.S. Department of Education website at www2.ed.gov/about/contacts/state/index.html or your nation's equivalent agency.

DO A LITTLE DETECTIVE WORK Some fake schools are clever and sophisticated. Some . . . less so. Here are a few simple first steps to take.

Call Them A sketchy-sounding voicemail during business hours, or a sleepy kid at 3 a.m. their time, is not a good sign.

Click Learning who owns the school's website can be a very rich source of information, especially if it conflicts with what is said by the school itself. www.whois.net is the place to do this.

Zoom In Check the school's address on Google Maps or Google Earth. If you find a mailbox rental store, a WeWork space, or (as we saw recently) a Holiday Inn, it's probably not a top-notch university.

ASK SOME QUESTIONS If you've done the basic checking noted above and you're still not really sure how legit a school is, asking the following questions can help clear things up.

How Many Students Are Enrolled? Quite a few schools seem reluctant to reveal these numbers. Sometimes it's because they're embarrassed about how large they are, as in, for example, the case of one less-than-wonderful school that at one time had more than 3,000 students and a faculty of five. Sometimes the school may be embarrassed about how small it is, as in the case of one highly advertised school that had impressive literature, very high tuition . . . and fewer than 50 students.

How Many Degrees Do You Award Each Year? As with number of students, the answer might be strangely high (as in, as long as

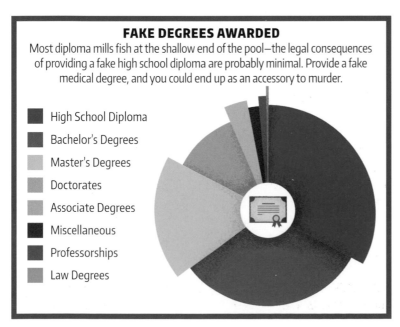

FAKE DEGREES AWARDED

Most diploma mills fish at the shallow end of the pool—the legal consequences of providing a fake high school diploma are probably minimal. Provide a fake medical degree, and you could end up as an accessory to murder.

- High School Diploma
- Bachelor's Degrees
- Master's Degrees
- Doctorates
- Associate Degrees
- Miscellaneous
- Professorships
- Law Degrees

IS THAT ONLINE SCHOOL BEING JUST A LITTLE TOO PUSHY IN OFFERING ITS "EDUCATIONAL SERVICES"? IF ITS REP SOUNDS MORE LIKE A HEAVY-HANDED SALESPERSON THAN AN EDUCATOR, YOU'RE PROBABLY DEALING WITH A DEGREE MILL.

KEY CONCEPT

WHAT YOU KNOW

One concept that questionable schools take advantage of is that of prior, or "life experience," learning. This can be a valid—if sometimes controversial or misused—tool. For example, let's say that an American student who takes four years of French at the university level can earn 24 credits. But what about a student who learns the language herself from online classes or while living in Paris? Many schools will give her some language-learning credit (although rarely that full 24 credits) once she's demonstrated her proficiency through written and oral exams. Some less-savory schools play to vanity by claiming that your life is so awesome that you've earned a degree just by living it! Or, more likely, that your years in business qualify you for an MBA with no extra study, something no legitimate school would suggest you do.

you send the check, we send the degree) or low (we just can't be bothered to fire up the printer).

LEARNING MORE ABOUT A SCHOOL Questions to ask would include: What is the size of the faculty? How many of these are full-time, part-time, and adjunct? Where did the faculty members earn their degrees? From which schools did the president, the dean, and other administrators earn theirs? While there's nothing wrong with faculty members earning degrees from their own school, once that number climbs to more than 25 percent, it starts sounding a little suspicious.

May I Have Contact Information for Some Recent Graduates? Ask for the names and email addresses of people who graduated in your field of study or who are in your local area. Most, but not all, reputable schools will supply this information.

May I Look at Some Coursework? An examination of master's theses and doctoral dissertations for the relevant degrees can often give a good idea of the quality of work expected of students and how well they rise to those expectations. Not every school will want to go through the hassle, or have obtained the proper permissions from students, but it's worth asking. Not everything is digitized; you may need to pay photocopying and shipping fees. Or visit the school—if you can afford to make the trip, it can be very revealing.

Will Your Degree Meet My Stated Needs? As noted earlier, there are several reasons you might seek a degree: state licensing or certification (for example, marriage therapy or primary education), requirements for graduate school admissions, qualifying for a new job, and so on. The school can't tell you with any authority whether the degree will get you a raise at your job or help you find a new one—and if it does, that's a red flag. But it should be able to speak to whether the degree will satisfy academic or professional requirements.

What Is Your Legal Status? If the school is located in the United States, ask about its status with both state and accrediting agencies. If it claims to be accredited, is it with an agency approved by the Department of Education or the Council for Higher Education Accreditation? If not, why? Does it plan to seek accreditation? If the school claims accreditation from a nation other than that where it is located, that's another red flag. Ask about this—usually it's because the accrediting nation has lax standards, but there could be a valid reason for the situation.

How Do You Handle Financial Aid? One of the necessary aspects of higher education in the United States is, of course, paying to enroll in the courses. If the college demands money up front, offers a pay-as-you-go system for its degree, or refuses to accept federal financial aid, these are serious warning signs to watch out for.—Dr. John Bear

THE TAKEAWAY

Many totally legitimate and academically rigorous degree programs are offered online. So are a lot of worthless scams. Here's how to tell the difference..

BASIC SECURITY	• Check the school out in a reputable college directory. • Only deal with accredited universities—and don't take it on faith that they're accredited. Check. • If it sounds too good to be true, it probably is. If the only "work" required for your degree is giving them your credit card, be very skeptical.
ADVANCED MEASURES	• Ask to speak with graduates of the program you're interested in. • Ask if you can "audit" some online courses to assess the instruction and student engagement. • Scope it out on Google Earth.
TINFOIL-HAT BRIGADE	• Visit the school in person to check out their facilities. • If you can't travel to the location, see if you can hire a local through TaskRabbit or Craigslist to play private detective and snap some images of the buildings and grounds.

ACCREDITATION
Standards and laws for accreditation vary by nation, state, or field. It's easy to say that a traditional university accredited by an official U.S. government–approved agency is the gold standard, but there are enough oddities, exceptions, and loopholes that things get confusing quickly.

In several nations, schools are authorized by national government ministries. Sometimes the process is rigorous, sometimes . . . less so. If academic integrity truly matters, check out those international schools carefully. If the school is somewhere unusual, ask plenty of questions.

In the United States, accrediting agencies are private companies that must gain recognition from the Department of Education after meeting fairly stringent standards. If a particular accrediting agency is not recognized, then you'll want to find out exactly why that is the case.

CHAPTER 10

SEX AND LOVE IN THE CYBER AGE

FROM CHASTE ONLINE DATING SITES TO ROUGH-AND-READY HOOKUP APPS, IT'S EASIER THAN EVER TO SWIPE YOUR WAY TO SEXY TIMES. BUT TODAY'S ONLINE ADVENTURES ALSO REQUIRE NEW SAFEGUARDS FOR YOUR DATA AND DEVICES.

Almost from day one, the internet has helped those with lonely hearts (or other body parts) make connections with others, as well as providing a wealth of erotic entertainment to the home viewer who might never have ventured into a shady adult bookstore in real life. As long as nothing illegal is taking place, and everyone involved is a consenting adult, you might ask what's the problem? Unfortunately, there are still bad guys out there who are very much aware that we tend to act impulsively when we're hot and bothered.

We might take more risks to get an immediate reward, such as clicking on a sexy link without thinking about the source. Sites promising illicit affairs and easy hookups could lead to personal data being compromised, and offers for easy money from online camming can prove to be a trap for the unwary.

From inadvertently downloading malware that cripples your computer along with those tempting sexy pictures, falling for a complicated catphishing scam, or having those compromising pictures or videos used against you, a wide range of financial, technical, and emotional risks abound for an unwitting user. This chapter will explore things to watch for and safety measures to take.

A TITILLATING HISTORY Collecting sexually sensitive material (kompromat), is a mainstay of espionage used to blackmail important individuals. Cold War–era Soviets skilled in this tradecraft were called "swallows" (women) and "ravens" (men). But if you have no shame, can you be blackmailed? Take the rather delightful example of Indonesian President Sukarno. On his way to Moscow, Sukarno fell for the friendly flight attendants and invited them back to his hotel for some hot action. The ladies, all trained Russian sparrows, gladly agreed knowing the scene would be secretly filmed for blackmail. After the deed was done, the KGB took Sukarno to a private movie theater where they showed him the film. Mistaking it as a gift, a delighted Sukarno asked for copies to be shown back home—where he insisted his country would consider him a hero for his prowess.

ONLINE DATING Whether you're looking for The One or simply a one-night stand, the internet is, to a greater and greater degree, becoming the place to go for those connections. Entire books and websites are dedicated to the basic etiquette of online dating as well as how to stay safe, whether you are having your encounters online or in person. Here are our top security tips.

Have a Conversation Most online sites and dating apps work pretty much the same: you view potential dates' pictures and profiles and then have a chance to message them. If they find you equally interesting, a chat ensues. This is a great way to get a feel for who someone is. Be cautious about divulging personal information too soon, but do take the time to get acquainted, and pay attention to cues. If someone pushes for your phone number, requests sexy pictures, or sends some to you unsolicited, these are major red flags. For better or worse, many potential dates will disqualify themselves early on through this sort of behavior.

Take It Offline Once you have a good feeling about a person you're chatting with, you'll probably want to suggest a meeting. It's best

SET UP A SEPARATE EMAIL TO USE JUST FOR DATING TO PROTECT YOUR PRIVACY IN CASE OF HACKS OR STALKERS.

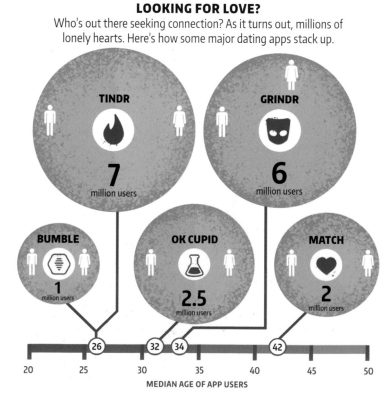

LOOKING FOR LOVE?
Who's out there seeking connection? As it turns out, millions of lonely hearts. Here's how some major dating apps stack up.

TINDR
7 million users

GRINDR
6 million users

BUMBLE
1 million users

OK CUPID
2.5 million users

MATCH
2 million users

26   32 34   42

20   25   30   35   40   45   50

MEDIAN AGE OF APP USERS

to make that meetup a safe and low-key one, such as getting together for a cup of coffee during the daytime in a public location. This is another tip that will disqualify some unsavory types right away. If the potential date comes up with excuses about why that's impossible, they may be a scammer looking to steal from you once they've gotten your confidence, or they might be a jerk who won't take "no" for an answer.

Be Cautious Before you meet up with your new acquaintance, tell a friend where you'll be and with whom, and arrange to contact your friend at a certain time. You can even inform the person you're meeting about your check-in plans as another test of their personality—and hopefully they'll be understanding. It might seem like overkill, especially if you have good instincts about someone, but this really just boils down to the "better safe than sorry" adage.

TRUE STORY

HOSTAGE BRIDES In Europe and the United States, some young women have been seduced over the web by ISIS members to become their brides. Recruited online with promises of the glory of supporting freedom fighters for a perfect Islamic state, young women connect via Twitter and Skype. They are directed to fly to Turkey, buy burner phones, and cross the border into Syria on their own.

Reports state that some brides fulfilled typical "wifely" roles (cooking, cleaning, and "comforting" the men), while others were sold as slaves, beaten, and raped. While it may seem unbelievable that smart teens could fall for this, let alone have the resources to make it to Syria by themselves, to an unworldly romantic, this might seem like a fairy-tale adventure, until it's too late.

In a lighter twist on this story, a recent news report told of some enterprising women who catphished ISIS soldiers, claiming that they wanted to travel to Syria and meet them but lacked the funds for airfare. Their potential suitors wired money, and the women subsequently laughed all the way to the bank.

KEY CONCEPT

CATPHISHING SCAMS Broadly used to describe the practice of creating a false online profile to deceive people looking for genuine relationships, catphishing is sometimes done by bored trolls looking to mess with peoples' minds. In more malicious cases, the catphisher may request sexy photos or ask for cash (often a "loan" so that they have an excuse to meet the victim in person). If you think you're being led on—if those photos seem just a little too staged, or if your new friend won't chat on the phone or meet in person—it may be catphishing. Many folks use stock photos or other peoples' profile images. Try dragging one of the pictures into Google image search and seeing what comes up. Your love interest may have also tried the same sweet words on other folks. Cut and paste a line or two from their come-ons, and see if anything shows up, perhaps in a post warning of this scammer.

RECKLESS REPORTING

During the 2016 Brazil Olympics, a journalist used various hookup apps—including Grindr, which enables its users to track each other by location—to look up gay Olympic athletes and out them. Aside from this being a serious breach of journalism ethics, outing others without their permission is disrespectful and, in some cases, can put their lives at risk. Some of those athletes may have been from nations where homosexuality is considered a crime—in some places, even punishable by the death penalty. If you're concerned about your own safety and privacy as a user, then you should probably avoid using these sorts of tracking apps. This means you'll have to find another way to meet new friends, but the lack of convenience is worth your safety.

STATS ON SEXTS

Who's sending all those salacious images and texts? As it turns out, quite a lot of people are getting digitally flirt.

HAVE SEXTED A PARTNER:
48% of women
45% of men

HAVE SEXTED A NON-PARTNER:
16% of women
25% of men

HAVE SEXTED A TOTAL STRANGER:
6% of women
10% of men

FLIRTING AND MORE Online communication often leads to flirting and sexy talk far more quickly than it would otherwise develop in person. There's something about the intimacy of chat (perhaps especially at home after a few glasses of wine, or when lonely and bored) that can make one get naughty fast. This can be good not-so-clean fun, but you shouldn't let those thrills make you vulnerable. Here are a few things to keep in mind before you share that saucy fantasy or racy image.

Safe Sexting You should think twice about sending racy messages over your standard text app, even if there's nothing inappropriate going on. After all, things might change, and a former partner might try to use revealing information to embarrass you. (You also don't want an X-rated message popping up on your lock screen if your phone happens to be on your desk at work where anyone can see it.) For anonymity, you can go with apps such as Kik or Wickr, which let you set up a user name rather than sharing your phone number. You can also use something like Facebook's messenger app, but be aware that the company has the right to read your messages and may use information in it to target ads to you.

Picture This Take extra caution when you're sending sexy images. You might accidentally forward a message or post it on social media. Even transient social media apps such as Snapchat aren't safe: People can still take screenshots, and the companies that host the servers for said software could potentially snoop.

Distinguishing marks such as tattoos, or background images or locations, can also be telling. A woman posted an image without her face on a Reddit forum, and one of the viewers on the forum recognized the bathroom as part of a dorm from a college in Florida. The viewer then posted the image to a sub-forum for that university, and the woman was identified and shamed. So you can't assume that distance equals anonymity—especially if there is anything in your environment that can be identified by viewers.

Just because you don't share a face pic online doesn't mean you're safe. A friend of mine felt deeply betrayed when she discovered that her ex-boyfriend had shared nude images of her with his friends while they were dating. The sexy photos didn't show her face, but the hurt and shame weren't made any better than that. They still knew who she was, after all.

TRUE STORY

EXTREME MEASURES Sometimes vindictive exes can be truly unhinged, as in the case of Michael and Tina (not their real names). After Michael broke up with Tina, she hired hackers to break into his Facebook and Gmail accounts. For four months, they sent harassing emails to his friends and family pretending to be Michael and making him out to be gay, crazy, and really into transgender prostitutes. Eventually, the hackers obtained naked pictures of him and posted them on his Facebook page. Michael was getting his MBA at the time and making business relationships to kick off his career. He estimated the cost of this reputation attack to be at least a few hundred thousand dollars in lost income. A digital forensics company traced the emails to outside the United States, but he could never prove who did it, so no charges could be brought. The odds of you experiencing such a thing are, of course, miniscule, but it's a reminder that anyone be hacked.

OVER 55 PERCENT OF PEOPLE WHO HAVE RECEIVED A SEXT SAY THEY'VE SHOWED IT TO AT LEAST ONE OTHER PERSON.

LOVE IN THE AGE OF CONNECTIVITY

A recently divorced investigator friend began online dating. He applied his online investigative techniques to the task and managed to weed out some seriously challenging dates who looked otherwise tempting. "The first thing I do when I see a prospective date," he said, "is search what I know from their profile: 'People named Melissa who live in Plano,' and 'Accounting.'" This almost always turns up, within a few minutes, an online profile—often with the same photo as used on the dating site. Then he runs his prospective date through PublicData, a website offering background and records checks. Within a few minutes, he can tell if the person has a criminal record, DWI arrests, or crazy social media accounts. This process, of course, is just as easy for her.

AMATEUR HOUR With webcams and streaming apps so prevalent, the line between sexting and pornography is increasingly blurred. Individuals who are comfortable with giving sexy shows online to a partner might easily decide that they might as well make a few dollars doing it for paying customers. For those who do, it can feel safe, because the interactions are livestreamed. That's a dangerous assumption, however, since it's easy for viewers to capture those live sessions on video, and those videos may show up anywhere. If you decide to go this route, just do it with your eyes open, knowing that family members, future employers, and others might see those videos at any time in the future.

"Eyes open" also means sober. Some websites and "real-life girls" operators are notorious for shooting footage of drunk women. Yes, they have to get releases for anything they later sell or air, but most women aren't in a sober frame of mind when it comes time to flash the camera. What might seem like innocent fun in the darkness of a club becomes a real problem in the light of day.

SAFE BROWSING Humans have been looking at sexy images for probably as long as we've been painting on cave walls. The availability of internet porn speaks to ancient desires—but with a whole host of modern considerations. Certainly not everyone views adult content on the internet, but many do, so without weighing in on the morality, here are some best practices and common concerns.

Financial Security Porn sites can't use the same credit card systems that other sites use (because many banks deny merchant accounts to adult service providers), so they don't have the same protections as other online retailers. While there are some systems that are well known and secure, it's recommended that you use anonymous prepaid credit cards for site access or purchases. Definitely don't use your regular debit or credit card. It's not uncommon for unscrupulous sites to add fraudulent charges on the assumption that consumers will be too embarrassed to complain to their bank if it means admitting what they'd been doing online. Don't fall for this—customer service reps at your credit card company won't mock you, and they're familiar with these scams.

Malicious Invaders Anything you download could have malicious software. Use protective software, special browsers, or a virtual machine (or a combination of these factors) to keep the integrity of your software. In addition, avoid clicking on ads while browsing, as so-called "malvertising" can infect your computer with one click.

Data Mining Data is constantly being collected about your browsing behaviors, even if you're using private browsing, and this data is then used to market to you. Sometimes this "marketing" comes in the form of viruses packaged in a link that hackers have determined you might find hard to resist, given the kind of adult media you prefer to view. And even cautious browsers can click impulsively if the promised images are intriguing enough. Again, protective software measures can help here, though they may not always reduce the data collection to zero entirely.

FUN FACT

DIGITAL DOMINATION The standard wisdom about dominatrices and their clientele is that the thrill of the transactional relationship has less to do with sex than with the exchange of power. Usually, these interactions involve a client paying the pro domme (the current term of art) for whippings, spankings, and such. A subset of this world is "financial domination," in which wealthy, powerful men (it's almost always men who are into this fetish) are turned on by letting a sexy someone take control of their finances for a little while (and a lot of money). Now, the brave new world of online sex has spawned a new way of flipping the power dynamic. A recent piece posted on Vice profiled Mistress Harley, a pioneer in the field of data domination. A tech-savvy online mistress, she charges clients for Skype sessions and ongoing interactions in which they may hand over their passwords, logins, banking information, and social media accounts to her, just for the thrill of it. She teases and torments them in pre-negotiated ways—examples might include blocking all porn on their computers until they beg and plead, using clients' Amazon accounts to ship sex toys to their workplaces, or posting dirty pictures to their Instagram or Facebook. This may be the ultimate expression of the truism that love (or something like it) makes fools of us all.

KILLER APP

GOING INCOGNITO Normally your browser keeps track of information to make browsing the web easier: your browsing history, cookies for login credentials, and a cache of downloaded images. When activating private browsing or incognito mode, these things are turned off. Dolphin is a Chrome extension that has an ad blocker incorporated and allows you to turn off all scripts, which protects you from self-executing malware. The Firefox browser, meanwhile, offers the ability to install add-ons such as NoScript and Adblock Plus, as well as a private browsing mode, to reduce the chances of your computer being affected by scripts and malware as well. This sort of software will ensure that your computer doesn't retain your browsing history, but it doesn't remove this information from the internet entirely. Your internet service provider and the sites can still monitor the traffic.

SEX AND LOVE

91 MILLION PEOPLE WORLDWIDE USE ONLINE DATING OR APPS

53% OF PEOPLE ADMIT TO LYING IN THEIR ONLINE PROFILES

AGE BREAKDOWN ON ONLINE DATING

18–24	27%
25–34	22%
35–44	21%
45–54	13%
55–64	12%
65+	3%

AVERAGE TINDER USER SPENDS 77 MINUTES ON THE APP EVERY DAY

TOP LIES ON ONLINE PROFILES

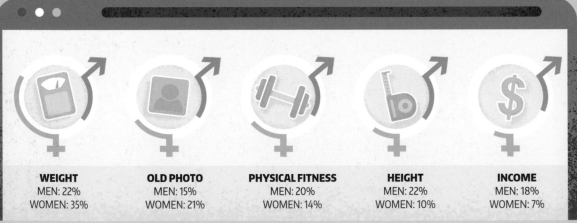

WEIGHT
MEN: 22%
WOMEN: 35%

OLD PHOTO
MEN: 15%
WOMEN: 21%

PHYSICAL FITNESS
MEN: 20%
WOMEN: 14%

HEIGHT
MEN: 22%
WOMEN: 10%

INCOME
MEN: 18%
WOMEN: 7%

22% Online daters who have friends to help create profiles

5 COUNTRIES WHERE THE SMALLEST PROPORTION OF WOMEN WATCH PORN

WORLDWIDE AVERAGE   WOMEN: 17%   MEN: 83%

JAPAN	GERMANY	EGYPT	UNITED KINGDOM	UNITED STATES
Women: 17%	Women: 17%	Women: 19%	Women: 22%	Women: 23%
Men: 83%	Men: 83%	Men: 81%	Men: 78%	Men: 77%

5 COUNTRIES WHERE THE GREATEST PROPORTION OF WOMEN WATCH PORN:

WORLDWIDE AVERAGE   WOMEN: 24%   MEN: 76%

PHILIPPINES	BRAZIL	INDIA	ARGENTINA	POLAND
Women: 35%	Women: 35%	Women: 30%	Women: 30%	Women: 29%
Men: 65%	Men: 65%	Men: 70%	Men: 70%	Men: 71%

25%

OF INTERNET SEARCH REQUESTS ARE PORN RELATED (68 MILLION). OF THOSE 68 MILLION, 116,000 ARE FOR CHILD PORN

35%

OF DOWNLOADS ARE PORN

8%

OF ALL EMAILS ARE PORN-RELATED (2.5 BILLION EMAILS)

20%

OF MEN ADMIT TO WATCHING PORN AT WORK; **13%** OF WOMEN DO

SECURITY BASICS

SECURE YOUR SELFIES

You might be wondering about your own personal flirty photos and whether you're safe from having private images stolen or being surreptitiously spied upon. Obviously the simplest way to avoid either is to not take such images or share them, but if you do share sexy photos, be sure the recipients know what you expect of them in terms of respecting your privacy; don't be afraid to ask for them to be deleted. If you want to keep that data, consider storing it on a USB or other portable drive disconnected from any computer and the internet, with the images encrypted or otherwise secured. As always, use strong passwords. And keep your webcam secured or cover up the lens with a sticker when you're not using it. Also avoid visiting any sketchy sites that might end up hijacking your webcam with malware.

REVENGE PORN One particularly disturbing result of the ease of taking and sharing of digital images is the rise of what's commonly known as "revenge porn," the public sharing of sexually graphic images without the consent of the subject, usually with the intent to harm. The standard case involves a vengeful ex posting images that might have been taken consensually at a happier time, but were never intended to be made public. After a breakup, the ex decides to try and hurt their former partner by publicly humiliating them or, even worse, opening them up to harassment, blackmail, or even attack. Malicious culprits can also post images online with doxxing, which means revealing personal information, such as someone's name, address, or work details, which can bring cyberstalking offline and physical harassment to the victim.

According to the Cyber Civil Rights Initiative (CCRI), 45 percent of revenge porn victims are stalked and harassed online by people who have viewed these images, with 30 percent also stalked in person. Some 77 percent of victims have faced social and occupational repercussions, and 48 percent say they have contemplated suicide. In the wake of suicides and lawsuits, 34 states have passed laws against revenge porn.

Some perpetrators post such images as part of a for-profit scheme. In 2010, for example, Hunter Moore, who started the site IsAnyoneUp?, became known as the revenge-porn king because the majority of the content on the site was nonconsensual sexual images—and not just of women. Claiming he was "just a businessman," he charged fees to remove the images from his site, netting an estimated $10,000 USD a month at the peak of popularity.

WHAT TO DO IF YOU ARE A VICTIM Revenge porn is illegal, and if you are victimized, there are laws and nonprofits that can help. Many places have laws on the books, and cases have been successfully prosecuted.

Document the Violation For sites to remove images, you must have the original in order to prove you own the copyright. Document the usage. Don't just save a link; take screenshots of the pages, especially if they are on 4chan, Reddit channels, or revenge-porn sites, or in Twitter direct messages. Be sure to include the URL bar or the name of the poster in the screenshot.

Identify the Perpetrator If you can identify the person who posted the information as the person you shared the image with in the first place, you can likely press legal charges.

Remove the Images Many sites will remove the images through a Digital Millennium Copyright Act (DMCA) takedown if you can prove you own the copyright. Other sites explicitly extort money from victims to remove the images. Document all communications: Save emails, take screenshots of text messages, make notes of any phone calls.

If you own the image—for example, you were the one who took the picture—you own the copyright and can ask Google and other search engines to remove it from search results through a DCMA takedown notice. This won't remove the image from the website where it is posted, but the image won't show up in search results.

To begin the DMCA takedown process, you will need to submit a report to each search engine for each instance. This is why it is important to take screenshots. The search engine will require that you:

- Give the URL of the website that is infringing on your copyright.
- Prove you own the copyright by attaching the original image to identify the copyrighted work If you don't have the original, you might not be able to prove you are the copyright owner.
- Sign sworn statements, and sign and date the submission.

Press Charges There are laws against revenge porn in thirty-four states as well as Washington, DC. Revenge porn laws fall under stalking and harassment, unlawful distribution of sexual images, disorderly conduct, violation and invasion of privacy, nonconsensual pornography, and unlawful dissemination of sensitive images.

T/F

YOUR WEBCAM CAN BE HACKED

TRUE The good thing is that this is a bit complicated. Before hackers can take over your webcam, they have to convince you to install malware. How do they do that? They could send you an email with a link to launch a script starting an installation process, or they can send the script hidden in a document, image, or video file. Once the malware is installed on your computer, it takes some skill on the hacker's part, but your cam can be compromised. How do you protect yourself from this scenario? Don't open or download unusual documents from people you don't know. Also, use a piece of tape or other webcam covering when it is not in use.

- - - - : : - - -

PRACTICING SAFE SOFTWARE If you want to view naughty images or video online, you can do so safely with a few precautions. Use a web browser that doesn't automatically run scripts, Java, Flash, or Adobe Reader—or add a plug-in such as NoScript, which lets you control the scripts run by each site you visit. And use an ad blocker—some ads can have malware and executables in them. This also helps to stop sites from redirecting you without your consent. You can also use browser plug-ins such as Ghostery to show what scripts are running in the background of every website. As you browse, check the link in your browser bar at the bottom of your browser to see where it is taking you. Stay away from sites that redirect your ride, and avoid the chance of infection by steering clear of downloads of any kind. You can still watch sexy streaming videos online.

KEEPING YOUR MACHINE CLEAN As with any online activity, use common sense when browsing sexy sites. Legit porn operators know their sites can be targeted for attacks and infiltrations, so they're incentivized to catch problems fast in order for their customers have a positive experience. Research which are the safest porn sites online, and ask open-minded friends for recommendations. Here are some added levels of security you should employ.

Limit Your Devices Whatever device you use, be sure it's secured. Update virus protection and run malware and spyware checks on a regular basis. Clear your caches, too; your phone could be stolen, or your cloud account could be hacked, revealing your browser history, so delete images and clear your trash on a regular basis. And consider keeping it to one device—some people choose to view porn only on certain devices to lower their risks. For example, only watch it on a computer with up-to-date security software.

Keep It out of the Office It should go without saying not to use work devices for porn or sexting, but news stories tell us that this advice hasn't gotten through to everyone. In fact, a recent UK study reported that 10 percent of office employees admitted to watching porn at work—and those are the ones who fessed up, so we can imagine the real numbers are much higher. Depending on where you work, this can be a firing offense—and even if it's not, you really don't want to have that conversation with HR, especially if you've also infected your company's network with malware.

Go Virtual A virtual machine is a separate environment that runs on top of your existing operating system. Running a virtual machine enables you to browse without putting your whole operating system at risk. If you download malware or a virus, it infects the virtual machine, not your whole computer, and when you end the virtual session, everything in it disappears, including that infection.

Pay Safely Legitimate free porn sites do exist, many with limited material and incentives for membership. If you pay for content, be careful who you give your information to. Some credit card companies will give you a virtual number; when used, it charges to your main account, but if stolen, it's easy to turn it off without canceling the basic card. You can also use reloadable prepaid cards.

THE TAKEAWAY

When looking for love on the internet, sending racy images to a special someone, or viewing racy images yourself, you're vulnerable. Practice safe sexting!

BASIC SECURITY

- Secure access to your devices (using PINs and lock codes) and use two-factor authentication.
- Use different email addresses and anonymous accounts for dating or hookups. Never use your work email for anything dating-related.
- Use a private SMS and voice call app to communicate with potential dates or hookups.

ADVANCED MEASURES

- Watch free porn, or use prepaid credit cards and stop subscription payments after you cancel.
- Use bitcoin for payment.
- Hide your face and anything that would make you identifiable when taking sexy pictures.

TINFOIL-HAT BRIGADE

- Use secure apps only for sensitive messages and set a timed message delete.
- Never send a photo or video that you wouldn't want to have made public if it goes astray.

SOME ADULT SITES ACCEPT PAYMENT IN BITCOIN, SO YOU DON'T HAVE TO DISCLOSE ANY BANK INFORMATION. FOR THE EXTRA-CAUTIOUS, THE ADDITIONAL STEPS MIGHT BE WORTH IT.

INTERNET VIGILANTES AND MOB RULES

THE INTERNET HAS BROUGHT US CLOSER TO EACH OTHER—BUT TECHNOLOGY DOESN'T MAKE MANY PEOPLE NICER TO EACH OTHER. LEARN HOW TO PROTECT YOURSELF FROM ONLINE JERKS AND FROM EVEN BECOMING ONE YOURSELF.

Technology has connected us in a global village, but that doesn't mean we can always speak to each other civilly. According to a recent study by the Pew Research Center, 40 percent of adult internet users have experienced online harassment directly, and 73 percent have seen it happen. The technology that connects us to others around the world can be used to torment—both on- and offline. Welcome to the dark world of trolling, doxxing, Anonymous ops, and online mobs.

The internet has decentralized everything, and harassment and mass protesting are no different. Online harassment can escalate to death or rape threats, prank calls to police departments, canceled speaking events, ruined careers—even driving some victims off the internet completely. Who are these trolls? Some want to entertain themselves or get off on manipulation, but the majority are bored teenagers who have turned to the internet to create their own drama, to get back at their friends, or win one-upmanship points. Having said that, the same technology can also be used for positive ends—stopping animal abuse, shutting down spammers, and protesting unfair internet laws.

Technology brings out the best and the worst in humanity. Learn how to protect yourself when the worst of humanity unleashes itself on the internet.

MEET YOUR INNER TROLL It used to be thought that mean people are born that way. But a new study from Stanford University suggests that, under the right conditions, anyone can be a troll.

The experiment exposed subjects to negative moods and/or comments and then asked participants to make their own comments. Those exposed to either the negative mood or comments were more likely to post negative statements; subjects exposed to both negative mood and comments were even more likely.

Negative comments can have serious emotional power—causing a downward spiral, with users returning to defend their statements and dig in their heels deeper. The negativity builds on itself and keeps growing. It's true: Just like laughter, trolling is contagious. Next time you're cranky and tempted to troll, pause, take a breath, then step away from the internet to cool off.

KEY CONCEPT

TROLLING The word "troll" conjures up images of a monstrous figure lurking under a bridge, but its origins are a different beast altogether. The word comes from the verb "to troll," which describes the fishing method of dragging a lure as bait. Internet trolls are similarly using "bait" when they post incendiary, hostile, and provocative information in order to lure others into having an argument with them. Trolls love to provoke people to get a reaction, and they also enjoy keeping the game going as long as possible, indulging in all of the anger and frustration they evoke. In internet-speak, lulz, derived from the slang term LOL (from "laugh out loud"), is laughter at the expense of others, a sort of modern schadenfreude. "Doing it for the lulz" means that trolls do what they do specifically so they can get an emotional rise out of their target.

BAD BEHAVIOR Trolling, harassment, and bullying create emotional distress and can lead to offline violence and real-world crimes, such as stalking or swatting. Trolls pop up in video game chats, review sites, on forums (especially 4chan, 7chan, and Reddit), and social media sites, including Twitter and Facebook. Trolls love all comment sections—in news stories and on YouTube, Tumblr, and even your blog if you've caught their attention. Companies have been trolled on Yelp, and individuals (including a White House spokesperson) have even been trolled on the payments platform Venmo. One thing is certain: Trolls are creative, and where they can troll, they will.

What They Do Online harassment includes sending nasty emails, sharing victims' personal information online, and calling for violence against targets. Many women even receive death and rape threats.

Stay Clear Trolls have no rules and will contradict themselves. They aren't logical, so don't bother trying to reason with them. They start arguments, post negative and shocking comments, give wrong information, and get people riled up. They're after angry reactions that keep the reprisals coming—they delight in creating chaos.

Trolls sometimes work together, posting targets in shared troll forums so that many of them go after the same target. These trolls get cred for participating in such ops.

GOOD TO KNOW

KNOW YOUR ENEMY Trolling takes a number of forms, including these common ones.

Dogpiling An internet cybermob descends on their target, trying to overwhelm, exhaust, and humiliate the victim.

Concern Trolling Someone gives "helpful" advice that's actually meant to belittle and demean the target.

Gaslighting The act of manipulating victims to make them doubt their own perception, memory, and sanity. If an abuser says it's not abuse, you are being gaslit.

Gish Galloping The objective in gish galloping involves a nonstop attempt to wear victims down, waste their time, and distract and sidetrack them. It's death by a thousand microcomments.

Impersonation Trolling As a way of discrediting targets, some trolls create fake social media accounts in their names and post provocative statements. Trolls might even take these false statements from the fake social media account and accuse the account holder of having made those statements.

Newbie Trolling There are always newcomers to online communities, and these folks can be taken advantage of by those sharing bad advice or giving misdirection.

Sea-Lioning These trolls join an online conversation and ask targets for their evidence. While generally civil, they question the facts and constantly challenge them—and then play the victim when their harassment is called out.

Shock Trolling Similar to radio DJ "shock jocks," these trolls go into a sensitive community (usually a religious or political forum) and stir up trouble by posting incendiary responses, images, or links.

HACKER HISTORY

ANONYMOUS ARISES
4chan, one of the most infamous forums on the internet, was started in 2003 by fifteen-year-old Christopher Poole. The notorious /b/ board, with an anarchic anything-goes attitude, is where many classic internet memes originated (such as Rickrolling and LOLcats) along with the hacktivist group Anonymous.

Legend says that Anonymous's first op was to save a cat from a teenage abuser who had posted YouTube videos of himself beating the animal. Forum members tracked down the poster, called the police, and had the cat rescued and the abuser arrested. The decentralized community went on to execute "ops" such as Project Chanology (targeting the Church of Scientology), DDoS attacks against other targets, and support of the Occupy movement. Anonymous derives its name from the anonymous nature of the 4chan /b/ board.

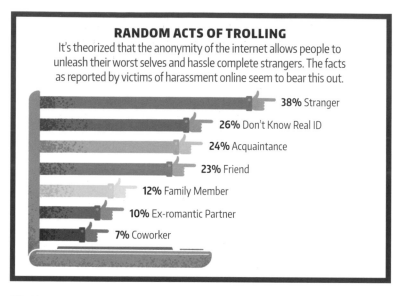

RANDOM ACTS OF TROLLING
It's theorized that the anonymity of the internet allows people to unleash their worst selves and hassle complete strangers. The facts as reported by victims of harassment online seem to bear this out.

- **38%** Stranger
- **26%** Don't Know Real ID
- **24%** Acquaintance
- **23%** Friend
- **12%** Family Member
- **10%** Ex-romantic Partner
- **7%** Coworker

THE ONE THING TROLLS WANT FROM THEIR TARGETS IS A REACTION . . . ANY REACTION AT ALL. BY DENYING THEM THIS SATISFACTION, YOU MAY WELL MAKE YOURSELF A MUCH LESS ATTRACTIVE TARGET.

KEEP CALM AND CLICK ON Trolls poke at you in order to evoke an emotional reaction. The most important thing to remember is "don't feed the troll." Sure, it's hard to control yourself when someone is pushing all your buttons, but that's what the cyberbully is after. Stop a moment, take your hands away from the keyboard, take a breath, and get up and go for a walk. If you don't respond, the troll will get bored and go after an easier target.

Online communities have dealt with trolls from the beginning of the internet, and many have developed rules and software to keep trolls out. Soft banning is a technique that basically hides troll posts from everyone except the troll. If no one sees the post, it won't get attention, so the troll gets bored and goes elsewhere.

You can't predict whether you will be the target of online harassment. Women are typically harassed more than men, but anyone can be targeted for any reason, including political or religious beliefs. If you end up the victim of a troll or online harassment, here are some helpful responses.

Don't Engage Do not respond or show emotion when provoked. If you show the slightest reaction, they will go in for the kill. They can be relentless, and sometimes when they don't get a response, they will escalate further to provoke you into responding.

Document Everything Take screenshots of incendiary texts, tweets, and comments immediately, as they can be removed. Keep a harassment diary. If it's too upsetting for you to do it yourself, have a friend take over for you.

Protect Yourself Use strong passwords, and turn on two-factor authentication to make it harder to hack your accounts. Remove your personal information from the internet to make it harder to be doxxed (see "Key Concept," at right).

Take Legal Action Contact the appropriate authorities. Make a police report, although officers may not be able to help much unless the harassment goes outside the internet.

Vent Safely Trolls delight in upsetting people on the internet; responding or complaining online reads them. Instead of venting online, talk to a friend or family member you trust about your frustration or anger. Get some exercise, spend time with friends or loved ones, get out in nature, take care of yourself—eat well, drink water, and meditate to let the anger go. If it gets really stressful, seek the help of a professional.

Shut Them Out Unfriend, block, mute, and report trolls. Most systems have ways to block and report unsavory behavior—Twitter, YouTube, and Facebook, for example, all have block and reporting capabilities.

Give No Comment If you are posting an article or blog post that you think will be controversial, turn off comments. If a troll is responding to a blog post on your site or a Facebook post you started, you should feel free to turn off commenting midstream or to just delete the offending comment.

Disappear from Sight As a last-ditch option, consider creating a new identity for the forum or site on which you have been trolled.

> ## KEY CONCEPT
>
> **DOXXING** Short for "documenting," doxxing is when your personal information, such as your legal name, your address, phone number, email address, or other data is posted online—along with an open invitation for others to harass you. If people know your name and address, they can take harassment offline and it can get violent.
>
> Prank calls to law enforcement about a false threat associated with a target's home is called "swatting." The goal here is to have armed police show up aiming guns at innocent people.
>
> Swatting is illegal and can result in jail time. Notably, twenty-one-year-old Mir Islam, pleaded guilty to cyberstalking, doxxing, and swatting over fifty celebrities and public officials. He was sentenced to twenty-four months in prison. And after introducing the Interstate Swatting Hoax Act of 2015, Massachusetts Representative Katherine Clark became the victim of swatting.

GAMERGATE

THE INSTIGATORS AND SUPPORTERS OF GAMERGATE CLAIM THAT IT'S REALLY JUST ABOUT ETHICS IN GAMING JOURNALISM. BUT A LOOK A THE HISTORY OF GAMERGATE SHOWS THAT IT'S ABOUT FAR MORE AND FAR WORSE.

Gamergate started in 2014 with a blog post carefully crafted to rile up gangs of online trolls. As intended, it incited a serious cybermob trolling operation. The attacks got so out of control that they garnered media attention not just in the gamer community but in the mainstream as well. Gamergate was by no means the first major trolling op, but it was arguably one the most effective based on the number of on- and offline harassment attempts it spawned.

The seed that kicked off Gamergate was a nasty post written by Eron Gjoni, the spurned ex-boyfriend of game developer Zoe Quinn. They couple had dated for five months. After they broke up, Gjoni decided to harass Quinn to get back at her for the breakup. So he wrote a 9,425-word blog intended to get the gamer/troll audience to harass Quinn.

In aiming his blog post at the young, white, middle-class males who make up the vast majority of the gaming community, Gjoni was able to spur an organized attack—and not just on his ex-girlfriend.

Mostly teens used to playing first-person shooter video games, these young men had begun to feel threatened by the growing diversity in the gaming field as more women and people of color started playing and making games, as well as criticizing existing games for their lack of diversity and their tendency to stereotype women as sex objects. Gjoni had hit a raw nerve in the white male–dominated gaming culture.

The post resulted in a firestorm of on- and offline

sexual harassment not just for Quinn but for scores of other women in the gaming community. Using the hashtag #GamerGate, trolls used the mask of philosophical disagreement to launch online attacks including negative comments and tweets, doxxing, disturbing Photoshopped images of the victims, inappropriate reports of the content they created, and rape and death threats. Anita Sarkeesian who produces Feminist Frequency, a website analyzing women's portrayal in pop culture, was another target. Brianna Wu, another developer, was doxxed and subsequently received death and rape threats. (Wu has gone on to start a legal

"QUINN WAS JUST THE FIRST WOMAN TARGETED."

defense fund and launched a campaign to run for Congress in 2018.) The same people who claimed they were focusing on "ethics" delivered bomb threats to organizations and conferences where any of their female targets were scheduled to speak or be given awards.

The cybermob was relentless in its attempts to humiliate its victims, ruin their reputations, and destroy their peace of mind.

Gamergate was a perfect storm of online harassment, sexism, and immaturity. At its core, Gamergate has truly been about the online sexual harassment of women in the games industry. It caused untold emotional damage to its victims, while the perpetrators hid behind anonymous accounts. And it's not the first case of cybermob online sexual harassment, though its massive collective effort was notable. In an earlier case, Kathy Sierra, a prominent user experience expert, received similar harassment and threats, causing her to leave the internet in 2007.

Unfortunately, Gamergate won't be the last case, either. The internet is a perfect medium for anonymous harassment. Anyone can be a target, so take steps to protect yourself from trolls: Don't react to their posts, take screenshots of messages and images, report the incidents to authorities, and use strong passwords on social media accounts. Hopefully any trolls will slink away. But perhaps the best way to end online harassment is to not do it yourself—and to discourage others from trolling as well.

LESSON LEARNED

Gamergate was truly a watershed moment in the history of online trolling and sexual harassment. For the first time, an individual was able to strongly motivate and manipulate a sizable community to do his dirty work for him. Once the first rock was thrown, it kicked off an avalanche that was impossible to stop. Others joined in for the lulz, new targets were identified, and the mob swelled further in scale and behavior. Gamergate gained national media attention when the online harassment of its targets escalated to doxxing; rape and death threats; and threats of bombings, shootings, and other violent acts planned for conferences where the various women were booked as speakers. These techniques exploded during the alt-right media coverage of the 2016 election. And to think that this all started out with a jilted lover wanting revenge on his ex when he was ditched.

TROLLING GENERALLY ARISES FROM A COMBINATION OF BOREDOM, DISTANCE, ANONYMITY, AND OPPORTUNITY. MAKE SURE YOU DON'T SLIP INTO TROLLING OUT OF BOREDOM OR ANGER.

MOBS OR VIGILANTES? So far we have focused on the bad things that happen when people team up anonymously on the internet. That's not always the way it works. Technology is value-agnostic, and the same tools, techniques, and tendencies that have been used for evil have done a fair amount of good. Welcome to the world of hacktivism and decentralized political power, where anyone with a computer can be an activist, and where trolling can be used for justice or in protest of injustice.

Project Chanology We discussed earlier the way in which the group now known as Anonymous started out of the /b/ random forum on 4chan. And while much of those activities were for the lulz, there was an equal desire to use the power for good. Anonymous took on Scientology in 2008 when a leaked video of Tom Cruise promoting Scientology was removed from YouTube. The op was called Project Chanology. What followed were multiple DDoS attacks, black faxes (wherein an image of a black page is sent to the recipient's fax machine so as to use up as much toner as possible, causing cost and annoyance), and staged physical protests. Since Scientology is known to intimidate and harass protesters, Anonymous protesters wore masks to underscore their message: "We are Anonymous. We are Legion. We do not forgive. We do not forget. Expect us." The mask they chose, which have now become synonymous with the movement is the "Guy

Fawkes" mask worn by anti-government protesters in the graphic novel and movie V for Vendetta.

Anonymous didn't stop with Scientology. Since the group is decentralized with no single leader, anyone can create an op, and anyone who wants to participate in the op joins. Anonymous has gone on to support to Occupy Wall Street and protest the KKK, ISIS, and the terrorists who conducted various attacks in Europe.

Trolling the Westboro Baptist Church The Westboro Baptist Church is known for its campaign of hate, displaying antigay signs at rallies and military funerals, as well as picketing and harassing other religions. After the church's leaders decided to turn its efforts toward harassing victims of the Sandy Hook school mass shooting in 2012, Anonymous focused its efforts on them. The Westboro Baptist Church website was quickly hacked and shut down, the social media accounts of several of its members were likewise compromised, and a charge was led in petitioning the government to declare the church a hate group.

Chasing Child Molesters In 2007, Canadian citizen Chris Forcand was arrested on multiple counts of soliciting a minor after several members of Anonymous engaged with him online pretending to be minors and collecting logs of the chats and inappropriate photos he sent. In 2011, Anonymous's efforts went further with Operation Darknet, whereby the group targeted and brought down forty hidden sites run by internet provider Freedom Hosting—all of them soliciting child pornography—and releasing the names of more than 1,500 users to the general public. In 2015, Operation Death Eater emerged, which is an effort to combat international pedophile groups.

Grassroots Advocacy Along with its larger-scale ops, Anonymous has also engaged in several smaller-scale or less-well-known activities, such as Operation Beast, which has targeted sites engaging in animal abuse; Operation Monsanto, which focuses on protests against the agriculture corporation; and Operation Safe Winter, a campaign working to create awareness of the difficulties faced by homeless people during cold winter months.

HACKTIVISM The internet has provided new opportunities for people to connect with each other—although not always in the best of ways—and one of those opportunities has been the ability to participate in social and political activism without having to leave one's home. Reputedly coined in 1994 by a member of the hacker group called the Cult of the Dead Cow, hacktivism is a combination of "hacking" and "activism" rooted in hacker culture and ethics. Those who are dedicated to a cause can use the internet in a subversive manner to promote social or political change. Most often the desire is to spread information and push for civil and human rights. Of course, due to the less-than-savory ways in which some have abused the internet, hacktivism has also had its unpleasant aspects, such as the Gamergate phenomenon (see pages 156–157 for more).

STAYING UNDER THE RADAR Do you know if your personal information—like your full legal name and your home address—is available online? Many sites have made a business model making this kind of information public. Others are looking to make it easy to create family trees. And on some sites, your information comes up with a simple search. Recently passed laws, however, require these sites to remove you from their databases if you request to opt out. Of course, there are companies, such as DeleteMe, looking to make a profit by doing this automatically for you. The reason you might want to remove your data from these databases is because this is where trolls get the information to doxx you and bring online harassment offline. Make it harder for trolls to find your information by removing as much of it as you can now.

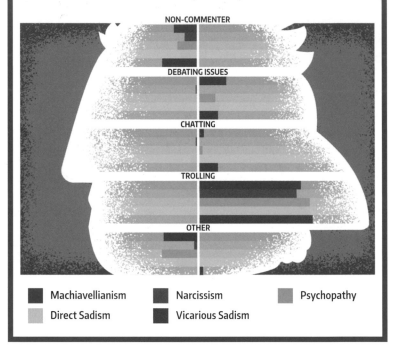

MEET YOUR INNER TROLL
Everyone has a little bit of a dark side, and trolling can bring out the worst in our natures. Here are some of the personality factors that make a troll.

NON-COMMENTER
DEBATING ISSUES
CHATTING
TROLLING
OTHER

- Machiavellianism
- Narcissism
- Psychopathy
- Direct Sadism
- Vicarious Sadism

WHAT MAKES A TROLL? Just as there will always be some unpleasant or troublesome people in society, ultimately the same is true on the internet. Trolling is unlikely to ever truly go away until humanity itself changes. A combination of distance, anonymity, and opportunity are a temptation that can sometimes bring out the worst in others, giving rise to their inner troll. But there may be a higher predisposition in some people. Psychological studies on sociopathy and antisocial behavior have presented the concept of the dark triad, a combination of three personality characteristics, that when combined, paint a dark picture of a malevolent individual.

The nefarious traits that are referred to by the dark triad are Machiavellianism, narcissistic behavior, and psychopathy, which, when combined, create a personality that has low empathy, a thrill-seeking nature, enjoys manipulating other people, and is focused on ego gratification. If you compare this to the activities of trolls—causing distress, doing it for the lulz, seeing victims as their source of entertainment, and "operations"—you can see how trolls fit this description. Given the opportunity, anyone has the potential to be a troll—but there are some people who may be born for it.

Remember the Good All this talk of trolls, vigilantes, 4chan, Anonymous, and Gamergate can of course paint a bleak picture that the internet is a foreboding place filled with potentially hostile figures waiting to bully or harass you for their own entertainment. But you should still keep in mind that there are also a lot of good people doing good things on the internet. By taking measures to avoid becoming a victim of trolling—and not becoming a troll yourself—you can guarantee that there will always be more good people than bad on the internet.—Heather Vescent

THE TAKEAWAY

Don't let your internet experience be ruined by a bunch of creepy kids and sad puppies. Here's how to avoid trolls or, if the worst happens, deal with them deftly.

BASIC SECURITY

- Block, mute, hide, and unsubscribe from troublemakers.
- Stay away from the comments section below online articles.
- Step away from social media if you start getting too worked up.
- Don't take even the worst attacks personally.

ADVANCED MEASURES

- Lock down the privacy settings on all social media.
- Use two-factor authentication for all accounts.
- Ask site admins if comments are moderated and stay away from any sites where they're not.

TINFOIL-HAT BRIGADE

- Only post under a pseudonym,;don't use your real name for any social media or forums.
- Create an email account you only use for social media (or other high-risk functions, such as online dating or gaming). Make the name nothing like your own and don't link it to anything else.
- Document online harassment and take it to the police.

DOING TROLLING RIGHT We can't, in all good conscience, actually tell you to go out there and troll others when you get bored. But we do know there are some people who can't resist the urge. But you should really know what you're getting into if you're going to let your inner troll out. If you're going to troll, do so with class. Use wit and refined thought as much as possible to point out the issue you've taken exception to. Pick a proper target—this means, as they say in comedy circles, "punching up." Choose an organization that has been problematic or harmful to innocent victims rather than belittling someone smaller and weaker ("punching down"). And always, always, be prepared for backlash. People out there will eventually decide they don't like how you've expressed yourself. Hey, we never said trolling was easy.

HACK THE WORLD

It wasn't too long ago that politicians and pundits were referring to the internet as our era's answer to the old Wild West—a lawless new frontier where fortunes could be made and reputations ruined. And the fact is, even if most internet searches are for cat videos and muffin recipes, plenty of sketchy stuff is still going on in the darker corners of cyberspace. Those shady back alleys provide shelter for good guys and bad from whistleblowers, political dissidents and those who leak secret documents to cyberspies, drug dealers, and global terrorists. Whether you're looking to bring down an oppressive regime, score a kilo of uncut heroin, or learn how to crash a nation's infrastructure, there's someone out there who's interested in helping. Can you trust them? Probably not. Red on before saddling up and heading off to conquer this new frontier.

THE DEEP DARK NET

YOU MAY HAVE HEARD MENTION OF THE DARKNET ON COP SHOWS OR IN NEWS STORIES ABOUT GUNS, DRUGS, AND MURDER FOR HIRE. IS IT REALLY SO DARK OUR THERE? SOMETIME IT IS—BUT NOT ALWAYS. LET'S GO FOR A DIVE AND SEE.

Aside from their rather evocative (and nearly meaningless) monikers, few people actually know what the "deep web" and "darknet" are—or understand the differences between the two. In this chapter, we'll look at what happens when you venture away from the brightly lit thoroughfares of the information superhighway and explore some of its shadier back alleyways. While in reality it's neither as glamorous nor as murderous as various crime shows on TV make it out to be, the fact remains that this aspect of the internet is a significant contributor to the world's economy—and to global criminal activity as well.

Although we can assume that a large number of the transactions on the darknet are relatively benign (although that depends on whether you consider stolen Netflix logins or off-brand Viagra to be harmless), many are not. As with any lawless wild frontier, you can find some real bad guys profiting off humanity's darker side. In this case, those bad guys might be dealing child pornography or heroin, selling hacked credit card accounts, or even involved in human trafficking. That said, some of the good guys out there can benefit from the same anonymity as well, particularly if they happen to be whistleblowers or individuals fighting oppressive regimes.

SHADY DEALINGS How much money is changing hands out there on the so-called darknet? More than you might think, although by their very nature, these sorts of transactions are obviously difficult to track and quantify. Reliable experts estimate that black-market transactions account for about 23 percent of the world's total goods and services. Think about that. If the global world product is almost $78 trillion USD, that means almost $18 trillion more is traded through the black market. In other words, almost a quarter of the world's transactions occur outside of the legitimate global exchanges. What are those transactions? Anyone who wants to make or receive untraceable (and untaxed) payments for anything, including illegal goods and services— from guns and drugs to hacking and stolen data—can find what they are looking for on the digital black market.

GOING DEEPER The first thing to know in getting your head around this topic is that the web is not the internet. While they may seem synonymous in daily life, the internet is far more than the World Wide Web. The internet is the entire global set of computers connected to a giant public network that shares certain rules for communicating various types of information. The most familiar of these are web pages and emails, but there are in fact many other things you'd never notice or care about that run over this same global network. Here's the basic breakdown.

The Surface Web Most of what you see online, from Facebook to eBay to Amazon to Twitter, is the surface web. It's made up of all the various public websites that share content, sell goods, or otherwise want to be easily found. They allow any guest to visit, and they invite search engines to index them so that users can find them through Google, Yahoo, and the like.

The Deep Web Millions of sites out there don't appear in search engines, often because they don't want to be found easily. These sites have no inbound links from any other site, and they block search engines. You can still visit them using a standard browser, but only if you have some other way of knowing the address, for example, if a link is sent in private email to a specific list of recipients. This type of arrangement is often used to share content, such as hacked data or child pornography, with a closed community that wants no outsiders.

The Darknet Simply put, the "darknet" is anything that cannot be accessed via a standard browser because it requires special software, and often special knowledge, to access. The darknet typically refers to sites on the Tor network (more on that on page 170) that look and feel just like regular sites but require a special Tor browser to view. The darknet, more broadly, includes other protocols and environments common users don't know about, such as IRC (internet relay chat) channels and I2P (Invisible Internet Project) networks. In addition, the darknet isn't indexed the way surface web sites are; virtually all of its sites' addresses must be shared instead of searched for, and not everyone out there will be keen on sharing.

The bottom line is that, for average users, the deep and dark web may seem alluring or sexy thanks to television. In reality, what you'll find there is malware, viruses, illegal content, and criminals ready to take advantage of the uninitiated. Unless you really know what you're doing, keep out.

A PRIVATE PLACE
Not everything on the darknet is illegal. In fact, it was originally designed by the U.S. government to let their operatives and analysts anonymously explore the farthest reaches of the global internet, in search of information. Privacy and anonymity are paramount to all users of the darknet, and many use it simply because they can communicate free from fear of online surveillance. Consider life in some countries where freedom of expression is not a right. Learning about, for example, government abuses by accessing foreign websites blocked by your nation is now safely possible using the darknet, as is expressing hopes and dreams—and, yes, buying a pair of counterfeit Versace sunglasses. When you consider the arbitrary nature of laws being an extension of the arbitrary power of government, the beauty of the darknet is clear.

KEY CONCEPT

SHOPPING IN THE DARK While it offers many legitimate uses to activists, whistleblowers, law enforcement, and political refugees, the darknet also supports an underground black-market economy that follows its own set of rules. Buyers and sellers have many reasons to trade outside normal open markets. These can include the following.

Illegal Goods Anyone looking for products or services that can't be sold in the open, such as drugs, stolen identity data, or weapons.

Anonymity These transactions can be done without records, allowing the buyer and seller to remain anonymous, with (theoretically) no paper trail or electronic footprint.

Price Controls Products that are subject to taxation, import duties, price controls, and other constraints—such as cigarettes or alcohol—are attractive to black-market profiteers. So are products such as Tide laundry detergent, Gillette Mach3 razors, Crest whitening strips, and other huge-markup common necessities.

Technology has made it easier for black-market buyers and sellers to safely connect and do business. In the constant back-and-forth between authorities and black markets, one black market is shut down, but another takes its place.

T/F

YOU CAN HIRE A HITMAN ON THE INTERNET

TRUE The most notorious online black market was Silk Road, operated by a hacker known as the Dread Pirate Roberts in 2011. It was an online marketplace functioning much like eBay or Amazon, except the offerings included guns, illegal drugs, hacking services, and even murder for hire. Silk Road was shut down by the FBI in November 2013, and Ross William Ulbricht, the Dread Pirate Roberts, is now serving a lifetime prison sentence. Not surprisingly, countless other illicit marketplaces have sprung up in its place. It's not entirely clear whether the many "hit for hire" services are pure fantasy or con men preying on the gullible. On the other hand, if they were as good as they say, how would we know?

IF YOU DO DE-
CIDE TO GO
EXPLORING
THE DARKNET,
DISABLE ALL
SCRIPTS IN
YOUR
BROWSER.
YOU NEVER
KNOW WHAT
MALWARE OUT
THERE COULD
BE CAPABLE
OF ATTACKING
YOUR
COMPUTER.

PEELING AWAY THE ONION The largest and best-known element of the darknet is the Tor network. Tor, which stands for "The Onion Router," was originally a project started by the U.S. Navy but has long since been turned over to a private nonprofit organization. The details are extremely technical, but, as an average user, you can think of the Tor network as three related technologies.

A Web Browser The Tor browser works like a normal web browser, but it routes users requests for web pages through the Tor network.

Safe Passage Tor anonymizes users' activity by stripping identifying data from page requests, then sending the requests through multiple encrypted transfers between volunteer-run computers all over the world that run special Tor software, letting them act as transit points. When a user with a Tor browser types in a standard web address, that request goes from the "normal" web into Tor, gets bounced through various intermediate relays, then reenters the normal web via a Tor "exit node" and arrives at the destination website. The site responds to the page request, and the content is sent back to the user by the Tor network through a similar process.

Hidden Websites Computers equipped with the right Tor software can also run websites (and other services, including IRC chat channels) accessible a via Tor browser. So users with a Tor browser are able to not only anonymize their browsing of the standard web but can see a whole second "web" (albeit relatively small) made up of sites ending in .ONION that can't be reached by a standard browser.

Software for browsing acting as a node or hosting a Tor website can be found at TORproject.org.

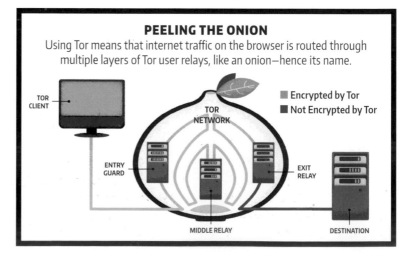

PEELING THE ONION
Using Tor means that internet traffic on the browser is routed through multiple layers of Tor user relays, like an onion—hence its name.

TOR CLIENT

TOR NETWORK

■ Encrypted by Tor
■ Not Encrypted by Tor

ENTRY GUARD

EXIT RELAY

MIDDLE RELAY

DESTINATION

UNDERCOVER OPS While criminals lurk in the shadowy recesses of the darknet, so too do law enforcement officers from around the world. A colleague in the UK complained that he'd bought too many knives and drugs to know what to do with. I spend countless hours surfing Tor sites looking for child pornography producers (distinct from distributors, who are often careless enough to use the surface web to exchange their unforgivable goods).

Shopping for Trouble The intrepid cyber security reporter Brian Krebs has taken on so many darknet criminals that his local police department had to come up with a special procedure when someone calls to report a violent crime at Krebs's house after SWAT teams, alerted by darknet thugs that there was a "hostage situation," had converged several times on his house. Krebs has also had criminals send heroin to his house, and then call authorities with the tip to search his mail.

"The darknet is where the bad guys are," a high-ranking federal agent told me. "We've got to get good at being there and looking like we belong there, because that's the only way we get into the kinds of conversations and relationships that will enable us to get leads and stop plots."

Reputation Matters This raises an important non-obvious point: In a world of anonymous users, reputation—based on common interests and sustained, consistent activity—is the only measure of trustworthiness. That's something that keeps academics busy.

Of course, reputation is how good reporters and journalists arrange to meet with sources to safely learn of corruption schemes and criminal gangs. The work done by these journalists is important in helping law enforcement and academic researchers understand new trends, and reporters can sometimes break stories based on darknet trading patterns. Brian Krebs, for example, broke the Target hack. Steve Ragan, another cyber reporter, has for years reported on criminal malware and hacktivist groups.

Check It Out—Carefully Who else is on the darknet? The morbidly curious—those who want to see whether you really can buy heroin and speed (you can) or hire a hit man (you can). The authors recommend you have a look around, if only to see for yourself the kinds of wares (and warez, aka pirated software) for sale, to educate yourselves about how the rest of this book isn't a bunch of people making stuff up. On the darknet, no one knows if you're a dog, but they do know if you have bitcoin (more on this digital currency in a bit).

FUN FACTS

WEIRD WARES The darknet may be a place to find guns, drugs, and hit men, but here are some of the stranger things you can buy on the DL.

Fake Coupons Grocers and snack companies have lost millions on everything from discounted cereal to free bags of chips.

Social Media Followers Wanna feel popular? For the low price of $25 USD, you can get 2,500 "followers" on Twitter.

Immortality You too can learn how to live forever! As long as you believe what the people selling the formula tell you.

Original Red Bull The energy drink now found worldwide was adapted from a more . . . potent formula originally sold in Thailand.

A New Identity Want to disappear? There are lots of guides on changing your identity. Plastic surgery not included.

144,000

TOTAL AMOUNT OF BITCOINS THE FBI SEIZED FROM ROSS ULBRICHT, THE ORIGINAL "DREAD PIRATE ROBERTS," FOUNDER OF SILK ROAD.

$12 MILLION

ESTIMATED VALUE OF BITCOINS LOST IN A "HEIST" WHEN EVOLUTION MARKET DISAPPEARED FROM THE DARKNET OVERNIGHT.

$100 MILLION

VALUE OF BITCOINS LOST WHEN SHEEP MARKETPLACE SHUT DOWN IN 2013, SCAMMING ITS MEMBERS.

$820K

VALUE OF BITCOINS THAT A FORMER SECRET SERVICE AGENT ADMITTED TO STEALING DURING THE SILK ROAD INVESTIGATION.

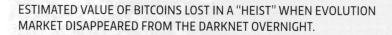

HERE'S A TIMELINE OF BITCOIN VALUE AT THE START OF EACH YEAR

2010	2011	2012	2013	2014	2015	2016	2017
.07	.30	5.10	13.66	839.62	273.98	914.68	433.38

BITCOIN VALUES HAVE FLUCTUATED BETWEEN $0.0001 AND $1,250.99 IN ITS LIFETIME THUS FAR.

30,000 UNIQUE SITES OR SERVICES (MAKING UP ABOUT 350,000 WEB PAGES) CAN BE FOUND ON THE DARKNET.

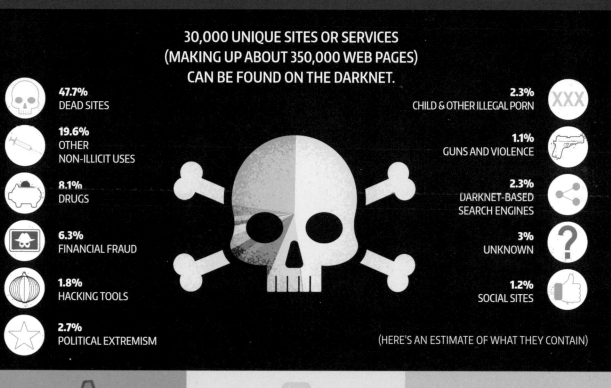

47.7% DEAD SITES

19.6% OTHER NON-ILLICIT USES

8.1% DRUGS

6.3% FINANCIAL FRAUD

1.8% HACKING TOOLS

2.7% POLITICAL EXTREMISM

2.3% CHILD & OTHER ILLEGAL PORN

1.1% GUNS AND VIOLENCE

2.3% DARKNET-BASED SEARCH ENGINES

3% UNKNOWN

1.2% SOCIAL SITES

(HERE'S AN ESTIMATE OF WHAT THEY CONTAIN)

3.4%

PERCENTAGE OF TRAFFIC THROUGH TOR USED FOR DARKNET ACTIVITY.

60 SECONDS

AMOUNT OF TIME IT TAKES TO SET UP A DARKNET BLACK-MARKET WEBSITE, USING STARTUP SOFTWARE DEEPIFY.

49

ESTIMATE NUMBER OF BLACK-MARKET WEBSITES CURRENTLY ACTIVE ON THE DARKNET.

$21.35 USD

AVERAGE COST FOR A NEW IDENTITY CARD

20% %: ESTIMATED FRACTION OF ALL WORLDWIDE DRUG SALES THROUGH DARKNET MARKETS IN 2014.

$1.2 BILLION

ESTIMATED AMOUNT OF MONEY MADE THROUGH SILK ROAD DURING ITS 28 MONTHS OF OPERATION.

DREAD PIRATE ROBERTS AND SILK ROAD

BITCOIN + EBAY-LIKE BLACK MARKETPLACE + AMATEUR TECHIE + VISIONS OF CHALLENGING THE U.S. GOVERNMENT = SILK ROAD. A STORY AS SHADY AS THE DARK MARKET.

The internet has enabled new business models that connect buyers and sellers from around the world for illegal transactions as well as legal ones. eBay was one of the first big online marketplaces, and it has carefully policed sellers to be sure no one is breaking the law. So where's an online shopper looking for something a little less conventional to go? It was only a matter of time before something arose to serve those interests.

The breakthrough for illicit online marketplaces came when bitcoin, a decentralized digital currency that works a lot like cash, was introduced. Word on the street was that bitcoin allowed you to make purchases online with perfect anonymity. This turned out not to be the case, but it's still a more stealthy way to operate than, say, using PayPal to purchase that shoulder-mounted grenade launcher you've had your eye on.

The first of the darknet markets, Silk Road was started in February 2011 by an anonymous self-taught administrator who later became known as the Dread Pirate Roberts, aka DPR. At the peak of Silk Road's popularity, it was estimated that the operation was bringing in $10K–$13K USD a month.

Silk Road emerged at a perfect time. It was like eBay but for black-market goods—mostly drugs, both illicit and pharmaceutical. As with eBay listings, sellers were rated, and there was an escrow to increase trust in the transactions.

There were certain illicit goods that were prohibited. DPR's philosophy was to hurt no one—

thus child porn and stolen data were not allowed. DPR saw Silk Road as a new brand that would challenge the government status quo.

In June 2011, Silk Road got major press on a number of tech blogs. This attracted more users—and also the attention of law enforcement agencies.

Task forces from the FBI and the DEA were formed to unmask and take down DPR. One DEA agent, Carl Mark Force IV, donned an alternate identity—that of a mid-level drug cartel player from the Dominican Republic—and

"IT WAS LIKE EBAY BUT FOR BLACK-MARKET GOODS."

reached out to DPR. They struck up a genuine friendship, but it was complicated because Force, as a member of the Baltimore Silk Road Task Force, was still trying to bring DPR in.

At one point, DPR asked Force to help him arrange the murder with DPR of a Silk Road employee who was cooperating with federal authorities. The task force managed to stage a convincing fake murder for hire.

The Silk Road bust came after months of chasing IP addresses and confirming site logins to DPR's account.

Chris Tarbell was the agent who eventually found Silk Road's IP address and traced it to a data center in Iceland. With a letter from the U.S. attorney, he walked out of the data center with a mirror of the drive, which was used to trace an IP address to Café Luna in San Francisco.

Ross William Ulbricht was the man arrested in October 2013 in San Francisco's Glen Park Public Library. All the bitcoins that were held on Silk Road servers were seized by the U.S. government and later sold in a series of auctions. Ulbricht was convicted of money laundering, computer hacking, and procuring murder. He is serving a life sentence without parole.

In a strange twist to the story, the undercover DEA agent Carl Force went rogue and was convicted of embezzling a small fortune of the seized bitcoins. He was sentenced to six and a half years in June 2015, including counts on money laundering and obstruction of justice. His colleague, Shaun Bridges, a former secret service agent, was also charged.

The Silk Road brand lives on, but it isn't controlled by Ulbricht. As of this printing, they're on Silk Road 3.0.

LESSON LEARNED

DARKNETS ALL THE WAY DOWN There have always been black markets, and they will never go away. Silk Road and the many sites that sprung up after its fall made buying illegal goods easy. As these sites became more popular, they also drew the eye of government. These black markets are lucrative, but most operators know they will eventually be busted. Buyers and sellers know this too and therefore keep minimal funds on the sites, in case they are seized. Criminals are both clever and greedy, and it seems like every new technology can be bent to nefarious uses. More tech-savvy criminals arise all the time, competing with laws and law enforcement—both sides aided by technology, which is itself value-agnostic. But how bad might it be, really, to have a clean, well-lit place for illicit transactions? For knowing the quality of the products sold? Some might say "Better the devil you know . . ."

KEEP YOUR IDENTITY OFF THE DARKNET

With the increasing numbers of site hacks, it's just a matter of time before your personal data is sold on the black market. There's not much you can do to protect yourself from someone else's site getting hacked—that's up to the site's security technology. But you can use good account password hygiene. Never reuse the same username/password combination on different sites. One of the first things hackers will do with freshly hacked data is automatically check the hacked username/password combos on numerous banking, social, and email websites. If you use the same username and password combination at other sites, hackers could get into your accounts on those sites, even though those sites were not hacked. Use a complex but easy to remember password combination.

LET THE BUYER BEWARE Until recently, the U.S. dollar, in cash, was the preferred currency for black-market transactions. Cash is untraceable and anonymous, but it's difficult to use cash for online commerce. Credit cards and other common payment methods leave a paper trail. That's why bitcoin is so ideal for shady shoppers. While as noted above they're not entirely anonymous, the use of bitcoin "tumblers" and anonymizing sites can obscure ones trail pretty well. Should you do this? There are legitimate reason to keep a light financial footprint, particularly if you subscribe to any number of concerns (or conspiracy theories, to unbelievers) about your government and its nosy ways.

Give It a (Careful) Go There are some reasons why a noncriminal user might consider these underground markets. For example, you might wish to rent access to the internet via another user's computer elsewhere in the world where, for example, Netflix shows first-run films not available to U.S. users, or you reside in a part of the world where freedom of expression is limited and you want to be able to communicate with others about civil rights or human rights issues.

Approach with Caution Small mistakes in your "operational security" can have massive consequences. It's not enough to be careful. You must be *very* careful and follow a strict set of procedures every time you enter and leave the darknet and even the deep web.

First off, don't use your own computer for this kind of exploration. In fact, you shouldn't even use a real-world one. Instead, download a bootable Linux image and always be sure to load your Tor sessions through that path. You should also load tools such as PGP (Pretty Good Privacy) encryption onto that bootable drive.

This setup will allow you to load your entire browsing session in the host computer's memory so that, when you finish, you restart the computer and there are no traces of the activity on your hard drive.

We don't recommend doing any shopping of course, but if you do get curious, don't just use Tor to access darknet

sites. For extra anonymity, you'll want to use an additional VPN (virtual private network) to completely anonymize your traffic.

All these techno-stealth measures may seem like a lot of work, but they're really the only way to access the darknet with any degree of security. And that's important once you think about the fact that almost everyone who gets caught doing something illicit gets caught because of security lapses. We will even go so far as to recommend using a clean laptop—completely devoid of any personal data or links to legitimate online accounts such as banking and thus dedicated only to your deep web and darknet adventures. You certainly don't want to be playing around in these neighborhoods with a computer that, if breached, would reveal a lot about your activities.

THE TAKEAWAY

The darknet is a fascinating place to spend a little time exploring, but dangers lurk everywhere, even if you're not doing anything illicit. Take reasonable precautions.

BASIC SECURITY

- Don't engage in any kind of illegal or questionable activity on the Internet.
- If you do any transactions on darknet sites, even perfectly legal ones, use encryption for everything.
- Disable all scripts ion your browser before logging on to Tor.

ADVANCED MEASURES

- Only use cryptocurrency for darknet transactions, and employ a tumbler to ensure optimal anonymity.
- Change usernames and passwords frequently.

TINFOIL-HAT BRIGADE

- Minimize coin kept in escrow to avoid losing it in a bust or heist.
- Use both Tor and a VPN to completely anonymize your traffic.
- Keep your data on a thumb drive so that you can erase all traces from your regular machine.

GOOD TO KNOW

THE FUTURE OF THE DARKNET Since the darknet has come into being, multiple changes have already taken place, and things will continue to change. Specific marketplaces will come and go—they're never going to go away entirely. Regulatory changes may influence what is sold on the black market, and some goods (such as marijuana, in places where it has been legalized) may transition to white markets. So what might we find on future black markets? In short: anything that is unregulated or highly regulated. This could be technology and drugs to augment the human body, government secrets, or new types of personal data, such as medical data collected by new consumer devices or household sensors. One thing is for certain: Future systems will be more secure than the ones that we have today—but future hackers will be more sophisticated as well.

WIKILEAKS AND WHISTLEBLOWERS

SOMETIMES WHEN CONFIDENTIAL INFORMATION IS LEAKED TO THE PRESS IT'S TO SERVE THE GREATER GOOD. SOMETIMES IT'S A DISASTER FOR HEALTH AND SAFETY. THE PROBLEM IS TELLING WHICH IS WHICH WHEN SECRETS SHOW UP ONLINE.

The internet allows all sorts of operatives and opportunities to get access to treasure troves of data that would have been unimaginable back in the days of Watergate, when actual burglars had to break into a bricks and mortar hotel to lift a few measly documents. These days, millions of pages of classified records can be liberated with a little stealth and skill.

This means that it's easier than ever for legitimate government agencies (or of course jack-booted thugs, depending on your perspective) to obtain all kinds of information from voice and written communications surveillance. This type of intelligence gathering traditionally falls into the realm of what is known as signals intelligence, or SIGINT. Our national intelligence agencies concerned with the capture and analysis of SIGINT have never had such an advantage, and they have risen to the task.

The flip side? What's good for the goose has turned out to be far better for the gander who wants to steal that SIGINT. This affects governments trying to protect information as well as businesses that are vulnerable to corporate and industrial espionage Ultimately "Information wants to be free," as the saying goes
. . . but at what price?

TRAITORS AND SCOUNDRELS In 1971, Daniel Ellsberg and Anthony Russo gave what became known as the "Pentagon Papers" to the *New York Times*. Ellsberg became that era's most celebrated (and/or despised) figure. The papers proved that presidents from Truman to Johnson lied about American involvement in Vietnam— and that Johnson had been bombing in secret. When the *Times* published, Ellsberg was heralded as a hero by antiwar protesters and excoriated as a traitor by those in government. This precedent set the stage for the WikiLeaks and Snowden affairs— among them, the understanding that newspapers can publish classified documents delivered to them (the Nixon administration had tried, in spectacularly poor fashion, to block publication), so long as they don't actively solicit their theft.

KEY CONCEPT

THE SECURITY TRIAD

The guiding principle of the security field is that you need three factors to ensure a secure system.

Confidentiality You simply must know that data stored on your system is protected against unintended or unauthorized access. That certainty is immensely complex to implement since, for example, Chelsea Manning (see page 181) was authorized to access—but not to copy and share—files.

Integrity The data's consistency, accuracy, and trustworthiness must be maintained over its entire life cycle, with contingencies for human error, server crashes, and viruses.

Availability The best data in the world is useless if you can't consistently and reliably access it, no matter what technology you're using.

WIKILEAKS AND SIMILAR SITES In the last several years, mentions of WikiLeaks and other associated websites and individuals have become more and more prevalent in media and conversations about security. But some people may still be a bit fuzzy on just who they are and what they do.

What Are They? WikiLeaks—as well as its wiser, more mature, and, from a policy perspective, more lastingly impactful older brother, Cryptome (as well as thousands of similar sites that have sprung up around the world from time to time)—provide varying levels of anonymity to those willing to disclose to the public information or data that they feel is of interest. This information can range from government documents (such as intelligence reports, diplomatic communiqués, program outlines, and planning descriptions), to insider stock trading records. Other examples include logs of computer network breaches, inside corporate policy documents not intended for public consumption (for example, internal pricing or policies on pharmaceutical distribution), customer records (such as the private customer data leaked by the hacker collective Anonymous on customers—including me—of the private intelligence firm Stratfor), naked photographs of celebrities, and anything else considered interesting or titillating.

Why Do They Exist? Reporters, muckrakers, short-sellers, investigators, opposition researchers, and suspicious spouses have always looked to insiders for these kinds of disclosures. The internet has simply made them easier to find.

What's Useful About Them? It can be argued quite well— and I do argue it—that our founding fathers had a hearty and healthy distrust of government, and they empowered the people to foment regularly this distrust through a vigorous and free press. If you believe that assertion, then you simply must believe that anything that empowers such a free press is, by definition, "useful." Regardless of your position on Edward Snowden, WikiLeaks, Chelsea Manning, or Daniel Ellsberg, that they are a part of public debate and discourse is ultimately better than why they are not.

What's the Downside? It is my personal belief that the kind of leaks inspired by Julian Assange—and committed by Chelsea Manning and especially by Edward Snowden—ultimately do more harm than good. When leaking is relatively easy (provided of course you have the access and the know-how), the kinds of thought and agonizing put into "what to leak" and "whether to leak it" exhibited by an Ellsberg or a Russo give way to the immediacy and the instant global celebrity attendant with the act of leaking, as we saw with the way Snowden gathered and disseminated some potentially deadly intelligence.

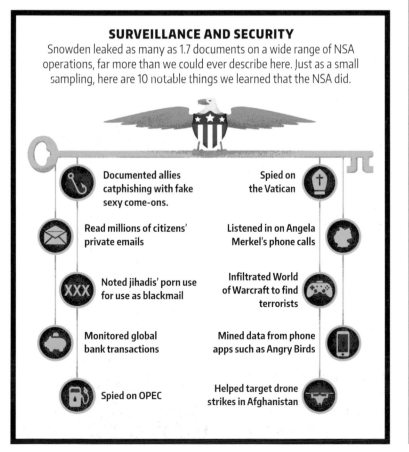

SURVEILLANCE AND SECURITY

Snowden leaked as many as 1.7 documents on a wide range of NSA operations, far more than we could ever describe here. Just as a small sampling, here are 10 notable things we learned that the NSA did.

- Documented allies catphishing with fake sexy come-ons.
- Read millions of citizens' private emails
- Noted jihadis' porn use for use as blackmail
- Monitored global bank transactions
- Spied on OPEC
- Spied on the Vatican
- Listened in on Angela Merkel's phone calls
- Infiltrated World of Warcraft to find terrorists
- Mined data from phone apps such as Angry Birds
- Helped target drone strikes in Afghanistan

EDWARD SNOWDEN
A former CIA employee, Edward Snowden rose to prominence in the public eye when, during his time as a contractor for the National Security Agency (NSA), he leaked over a million classified U.S. government documents to journalists. In June 2013, the U.S. Department of Justice charged him with espionage and theft of government property. He subsequently fled to Russia, where he currently has been given asylum until 2020. Snowden has claimed that he leaked the classified documents because he felt their contents were unconstitutional, and that he had become disillusioned with his government. As he has become a subject of much-heated debate in security circles and the media, Snowden has been hailed as a hero and a patriot by some, and reviled as a dissident and traitor by others.

STEALING FROM THE NSA The first question one might well ask when looking at the case of Edward Snowden is, why was it so easy for him to steal info? Well, it wasn't exactly "easy," but it wasn't as hard as it would have been in another era. Thinking about the key concept of confidentiality, and how difficult it is to implement it, things get harder at the level of systems administrator—someone entrusted to have access to a lot of systems for the purpose of making it work in the desired fashion.

The hardest thing to do on a network is classify data. Humans tend to overclassify data, and it's particularly difficult to classify the information that is already on the network after it's there (*ex post facto* classification), as opposed to classification of data that is being newly created.

The Human Factor In an intelligence agency, we begin with the premise that everything within its walls is already important enough to be secret—the fact that they collect it at all speaks to its importance. But even if that weren't true, even if Snowden worked at, say, an insurance company, where each and every fact had been accurately classified, the issue would have been the same: He still would have needed to be able to poke around the systems and ensure that all the whatsits were in the right place. That's because he was a systems administrator with administrative access, also known as super-user access. Typically, by the time you get to be a network administrator on a network that carries classified information, you've been checked out pretty thoroughly, passed lots of tests—including polygraph examinations in which you're quizzed about your "lifestyle" (this is, essentially, a government euphemism for your sex life)—and generally thought to be "squared away."

I spoke with a former systems administrator with Top Secret/Secure Compartmentalized Information (TS/SCI) clearance about this. The system Snowden stole from is at the highest classification level. In order to get access to the system, one would have to have the TS/SCI background check and the lifestyle polygraph mentioned above. As we said, Snowden was a systems administrator, so by the nature of his job, he would have had extensive access across the network; getting access to the data was fairly trivial for him at that point.

Getting the Goods The actual difficulty would have been getting the data out of the building. On TS/SCI systems, all removable media capabilities (like USB, CD, and DVD) are disabled, so there is no means for regular people to get information off the systems. There is an exception, though, for systems administrators. Within a sysadmin shop, there are typically one or two systems that have the ability to write CDs or take a USB drive. Sometimes, there are completely legitimate reasons why one would have to transport data between non-networked systems. For Snowden, the theft was not rocket science but a matter of abusing his trusted position.

Two things could have been implemented but weren't (and probably are now) to stop this: better internal data segmentation and two-person integrity controls (TPI) for systems administrators who want to use removable media—think of nuclear keys and you get the idea. It's a pain in the ass and it's inelegant, but it does work.

The fact is, getting that kind of access, despite how easy it may have looked, wasn't easy. It may have taken him five minutes to steal the data, but it took him years to know which data to steal and to be placed in the position that enabled him to steal it.

GOOD TO KNOW

BAD PRACTICE It's easy to look at the glamorous, freedom-loving aspect of whistleblowing without seeing the potential dark side or unintended consequences. In addition to facts and figures, Snowden's leaked documents also revealed the methods and capabilities of programs used by multiple governments to monitor covert communications on the internet—including methods to monitor those involved in child sex trafficking. Once these programs were revealed, these kidnapping, slaving criminals changed their tactics, forcing international law enforcement to find new ways to intercept and decode these transmissions. On that basis, forgive me if I don't refer to Snowden as a hero.

KEY CONCEPT

BLOWING A WHISTLE Not all informants are created equally. A "whistleblower" and a "leaker" are actually two separate types of individuals. In the case of the former, these are typically dutiful people who happen to discover something that is illegal or unethical, and then try to report the problem through the proper internal mechanisms; when they fail or are unable to do so (there can be a variety of reasons for failing), they report the wrongdoing to an external source but limit their reporting to only what has gone wrong. A leaker, meanwhile, can be considered someone who, whether out of carelessness or a desire to seek fame, avoids or ignores the standard channels followed by a whistleblower and instead disseminates the information in a less-conscientious fashion, without making much effort to do so discretely or with regard to the repercussions.

**JULIAN ASSANGE AND
WIKILEAKS** Australian-
born journalist and
publisher Julian Assange
is the cocreator and
director of WikiLeaks,
which publishes leaked
sensitive documents.
WikiLeaks has existed
since 2006 but came into
prominence as a result
of the documents leaked
by Chelsea Manning.
By 2015, WikiLeaks had
published more than ten
million of what Assange
describes as "the
world's most persecuted
documents." As with
other key players,
Assange has been called
a hero, traitor, and
opportunist. In 2010,
Assange visited Sweden,
where he became the
subject of sexual assault
allegations. He was
allowed to leave, but
later Sweden asked for
him to be extradited.
He has spent the last
several years living in
the Ecuadorian embassy
in London. In 2017,
WikiLeaks published
a trove of CIA hacking
documents said to be the
largest ever.

WHY AND HOW INFORMATION GETS OUT One key difference
between a Daniel Ellsberg, an Edward Snowden, and a Julian Assange
is that Snowden and the like are individual actors, whereas Assange's
WikiLeaks is a clearinghouse—a brokerage of information, if you will.
And thus, their methods, their motives, and their reception by media
and security experts vary.

Snowden's Motivation As an individual operator, Snowden's claims as
to why he did what he did are quite divisive. One side sees a freedom
fighter: a man truly dedicated to the idea that his government had
run amok, conducting mass surveillance of literally every adult in the
United States and Europe through extensive monitoring of a wide
range of technologies. To this group, giving up his six-figure income
and an arguably cushy life in Hawaii for a one-bedroom flat near the
Maryina Roshcha District is further proof of Snowden's dedication to
these fundamental propositions.

The other side sees agencies working to capture legitimate signals
intelligence of the sort that has been their remit since SIGINT was
the lynchpin on which the Allies achieved victory in World War II.
This intelligence is a well-defined and anonymized mass collection of
information that is freely available: call data records of numbers and
call metadata inside the United States, and call record and content
outside the United States joined only with properly defined and
legally sanctioned oversight. Along with traditional signals targets,
this intel is sought by every country on Earth for the same reason:
the protection of life.

Your personal beliefs about Snowden probably depend a lot on
what you do for a living, whether you've served in the military, and
your general political stance. And the "truth" probably lies in the
middle. From what we have seen, there appear to have been some
terrible abuses in the U.S. system of checks and balances, especially
when it comes to the Foreign Intelligence Surveillance Act of 1978
and its court. Many of the programs were described to the world by
journalists who admittedly knew nothing of intelligence, surveillance,
or even encryption before Snowden quite literally dropped the
materials into their hands (saying, cynically, that as journalists they
would know best how to release the information).

Assange and the Profit Motive Where Snowden might claim to
be inspired by Ellsberg, Assange sought to influence and provoke
leaks by people like Chelsea Manning. In that sense, Assange's
WikiLeaks behaves in a manner that is similar to an intelligence
service: Assange and his associates act as officers, who seek agents

in various positions of authority in governmental office to provide them with intelligence. The agents may turn over the intelligence for a range of reasons that they believe justify their actions, which may be anything from misplaced patriotism to revenge to idealism (WikiLeaks is not known to pay for leaks).

Depending on your viewpoint, this may be "better" or "worse" than traditional espionage, but espionage it is. Chelsea Manning's turning over of the documents that would ultimately make up the trove of diplomatic cables released by WikiLeaks presented the United States with a situation in which the entire world became privy in one fell swoop to the most intimate minutiae of its diplomatic communications—petty, trivial, damning, controversial, telling . . . It was an intelligence bonanza for any nation intent on understanding how the United States does this kind of thing.

GOOD TO KNOW

THE TOP THREE LEAKS Here are the most interesting, extensive, or devastating leaks in the history of modern intelligence breaches.

LEAK	YEAR	NATURE OF LEAK
PENTAGON PAPERS	1971	Daniel Ellsberg, then a U.S. military analyst, released documents to the *New York Times*, detailing more than twenty years of the government's involvement in Vietnam.
EDWARD SNOWDEN	2013	Leaked over a million documents relating to U.S. security policies and practices.
OFFICE OF PERSONNEL MANAGEMENT	2015	The United States Office of Personnel Management announced that 21.5 million records had been breached, consisting of all the personnel files on applicants for national security clearance.

KEY PLAYER

THOMAS DRAKE An NSA employee, Drake worked on "Trailblazer," a $1.2 billion USD data-mining program he felt intruded on civil rights and national treasure. After years of trying to report internally, he went public. He was arrested in 2010 and charged aggressively with willful retention of national defense information, obstructing justice, and making false statements. Drake spent much of the time awaiting trial (he refused to admit wrongdoing or to plea-bargain) working as a Genius at an Apple store. The government dropped all charges, in return for a guilty plea to misuse of an agency computer. The judge howled that the government mishandled the entire affair, saying it had destroyed Drake's career, taken his salary, and his pension. Drake got a year of probation and community service. Today, Drake is looked upon by many in the government as a genuine (if misguided and wrong) "whistleblower."

INTELLIGENCE AND DATA COLLECTION

For all the glamour that spy movies give it, intelligence is simply data that has been collected and then analyzed for a purpose. If you hide your lingerie in the top drawer of your dresser, and your child says that he's seen your sexy lingerie, you can conclude that your child has been in the top drawer of your dresser. There's certain intelligence in both WikiLeaks and in the stolen Snowden documents from which a foreign intelligence service can deduce or otherwise conclude our sources and methods. By giving adversaries insight into these, Snowden allowed them to close pathways of information collection. If they can conclude the sources, the sources are endangered. Arguably, others must presumably be placed at risk in order to establish new means of collections.

WHY IS IT SO HARD TO PROTECT INFORMATION? Data classification is highly complex, and because it is contextual, machines just aren't any good at it. The classic example of AOL locking people out of a breast cancer support group because AOL technicians tried to block the word "breast" may or may not be true, but it's certainly a good example of the difficulty faced by machines trying to discern the nuance of words in context. Why are chicken breasts fine but bare breasts racy? Even for humans, the process is difficult, because the context even of certain phrases can raise issues.

Technical Issues Data theft technology, which is referred to by the industry as data loss prevention (DLP), is fairly complex, but it's still very rudimentary in terms of intelligence. DLP is best at strings of defined lengths—credit card, social security, and account numbers are easiest to detect. But even within those, we have tremendous variation. For example, consider how you write phone numbers: 888/235-1212; 888-235-1212; 888.235.1212; (888)235-1212; (888) 235-1212; 888) 235-1212. These are all ways of writing the same thing. Now do it with words. How to compare all these variants in real time, as someone's trying to send an email and you're trying to scan it and determine whether the email contains something sensitive before the email goes out the door? Well, the trick is to buffer everything, truncate and stem all the words and phrases, remove all the extraneous characters, then hash everything, then compare hashes. It's faster. This can be done in an amazingly small amount of time. But it's still nowhere near foolproof; what if the file is encrypted?

The Enemy Within What if, as we've just discussed, the data thief is your system administrator? The fact is, catching data thieves is very hard unless you've classified it all very well in advance, limited access to it, removed the removable media options, and limited the ways to get data off your network. For most companies, that's not commercially feasible.

Bottom line? Data classification is very difficult. Businesses should pay close attention to how they classify and provide access to important data, and how people can get access to it. And data theft is incredibly hard to stop.

THE TAKEAWAY
What does government security and international espionage have to do with you? The lessons learned here are more helpful than you might think.

BASIC SECURITY
- Classify your data in two categories: public and private. Make sure to keep those records separate!
- Treat employees well. This is always a good idea, but particularly relevant if potentially disgruntled workers have access to classified information.

ADVANCED MEASURES
- Encrypt your private data whenever you email it and wherever you store it. And be incredibly careful about who has the keys.
- Destroy all data you don't need, regularly.

TINFOIL-HAT BRIGADE
- Use DLP software to detect data leaving your business.
- Go beyond the basics and classify documents and emails to understand when sensitive information may be leaving your network, and then speak with or take punitive action against employees who break your policies.

*

IF YOU SHOULD HAPPEN TO FIND YOURSELF IN A WHISTLE-BLOWING SITUATION, REVIEW YOUR LEGAL RIGHTS AND THE PROPER CHANNELS FOR REPORTING THE ISSUE TO AVOID BECOMING A LEAKER.

INTERNATIONAL CYBER SECURITY

THE DANGER OF A FOREIGN GOVERNMENTS LAUNCHING A CYBER ATTACK HAS BEEN A SECURITY ISSUE FOR SOME TIME, BUT NOT UNTIL 2016 WHEN RUSSIA HACKED THE AMERICAN PRESIDENTIAL ELECTION DID IT HIT THE FRONT PAGES.

We close this book by taking a look at cybersecurity on a global scale. Know who's already doing that? Every government on Earth, and each has been at it for a very long time. Nation-states consider "cyber" to be a key area of operations. It's where they communicate, spy, command, and control—and sometimes, where they attack.

Cyberspying against the United States became so problematic by 2011 that the military changed its policy on cyberattacks to "equivalency"—essentially, online attacks are now viewed just like physical ones. An unnamed military source told the *Wall Street Journal* that "If you shut down our power grid, maybe we will put a missile down one of your smokestacks." It was as a clear warning to Chinese and Russian hackers, the latter of whom had recently used cyber attacks to turn off the lights in Estonia, and then again in Georgia, as precursors to invasion.

Nations have a number of ways to rattle their cyber sabers. At the low end of aggression is intellectual-property theft and piracy. At the high end is the notion of crashing another nation's infrastructure or hacking its military. And what about non-nation-states doing such things? Could a teenage hacker really start a world war?

CUCKOO FOR CYBER SPIES Although cyberterrorism has been fairly big news as of late, it's actually far from a new concept. The contemporary cyber espionage era started up pretty much as soon as computers were first connected to public networks. For a real thriller of a hacker story, try reading Clifford Stoll's 1989 classic *The Cuckoo's Egg*. This nonfiction page-turner tells the incredible but true-to-life tale of how a systems administrator doggedly chased down a 75-cent UNIX-system accounting error, which unfolded into ultimately taking down a complex, KGB-funded cyber operation in 1986 that was seeking to steal U.S. military secrets through the fledgling internet. Stoll solved the case himself in spite of a series of infuriatingly dense FBI agents straight out of Central Casting, along with a cast of dozens of square-jawed spooks, each one of them more impossibly unhelpful than the last.

KEY CONCEPT

TRADE SECRETS

Intellectual property, or IP, is how businesses turn ideas into money—maybe a new fabrication process or the ingredients that make up their secret sauce. Here are key things companies keep in their IP portfolios.

Business Processes A closely guarded list of materials used to make products.

Bill of Materials A list used to make products, and closely guarded by manufacturers.

Software Firmware and apps are included. This is the core of many firms' IP portfolios and what makes a hunk of plastic into a beloved consumer electronic device.

Road Map The list of product features and functions that a company plans to introduce.

R&D The research and development of new products and features.

STEALING OUR GOOD STUFF So, what is intellectual property (IP) anyway? The legal definition is "creations of the mind, such as inventions; literary and artistic works; designs; and symbols, names, and images used in commerce." The term covers everything from art and music to apps and code. Theft of IP isn't super glamorous, so most cases, even multibillion-dollar ones, don't make the news outside of the business pages. The U.S. government recently estimated that cybertheft of intellectual property costs the economy $300 billion USD a year. If you find yourself wondering why cyber espionage is so prevalent, it's simple: It is substantially cheaper and faster to steal stuff than it is to build it from scratch.

"The only adversary one needs to worry about," says David Etue, from cyber security firm Rapid7, "is the one who figures out that he can steal for $2 million what it takes you $2 billion to research and develop." Etue is right. It's much easier for foreign government-controlled companies to simply steal their way to success than it is to build it through R&D.

Industrial Espionage It's not just commercial IP that gets ripped off. Drug trials, oil and fuel formulations, and other industrial secrets are in great demand. And it's not just the Chinese and the Russians doing the dirty work. Industrial espionage is top of the pops in France, as well as in many other nations. What a lot of people may not realize is that, to China and Russia, commercial adversaries count as targets for government espionage. Chinese and Russian companies are often owned by the government, so interference

THE COST OF CYBER THEFT

The thing about IP is that it frequently forms the core of a company's identity. A stolen computer can be replaced, stolen money can be recouped. A cyber breach of this kind is more like identity theft on a grand scale, and the real and intangible costs can be staggering.

VISIBLE COSTS OF IP THEFT
- Investigations
- Need to Notify Customers
- Monitoring Customer Security Post-Breach
- Regulatory Compliance Issues
- PR to Combat Negative Publicity
- Upgrade Cyber Security & Training
- Lawyers' Fees, Other Legal Costs

HIDDEN COSTS OF IP THEFT
- High Insurance Premiums
- Lower Credit Rating
- Lost Productivity & Low Morale
- Lost Customers
- Loss of Potential Future Business
- Reputation and Value of Brand Suffer
- R&D Time and Investment Wasted

from government or military hacking groups against American competitors is seen not a business chicanery but a matter of national security. It's not about the money, per se. It's about securing the future—especially in the realms of critical infrastructure, energy, medicine, and finance. It's business as usual.

National Security That's why, when the United States began shouting its protestations about Russian involvement in the 2016 hack of the Democratic National Committee, none other than Shawn Henry—the former assistant executive director of the FBI, for which he had largely established its cyber practice, and the man who led the investigation into the DNC hack for CrowdStrike—spoke out in the press about it in no uncertain terms: "It's the job of every foreign intelligence service to collect intelligence against their adversaries," he told the *Washington Post*. For the Chinese and Russians, commercial secrets and commercial organizations are considered legitimate nation-state adversaries.

CASE STUDY

HOW THEFT HURTS BUSINESS The poster child for nation-state intellectual property theft was American Superconductor Corp (AMSC) doing business with its Chinese partner, Sinovel Wind Group. AMSC made the software that drove Sinovel's turbines, and everything was just ducky until an AMSC engineer was on a Sinovel turbine one day and noticed that the firmware was, well, different. In fact, Sinovel had completely replaced the AMSC firmware with its own version, based on IP it had stolen from AMSC. Thus began the best-known case in the United States of theft by a Chinese company of an American firm's intellectual property. AMSC says it lost $1.2 billion USD, because Sinovel accounted for 80 percent of AMSC's revenues (making this case an object lesson on customer diversity as well as Chinese hacking).

You might ask why you should care if some random company gets ripped off. Simple answer: Who do you think they're going to pass those costs down to? Yep, the old end-user, a.k.a. you.

SECURITY BASIC

BORDER SECURITY A relatively new concern in 2017 was searches of electronic devices by U.S. Customs and Border Protection. The law allows these searches, but they are still rare—in 2016, there were 390 million crossings and 24,000 searches. Still, if you don't want Uncle Sam plowing through your hard drive, power down devices fully before crossing borders (cold boot security is often stronger than when merely suspended or locked) and minimize the amount and sensitivity of data and equipment you transport across borders. Be aware that citizens cannot be denied entry but can be detained briefly for questioning. Under no circumstances should you lie to CBP officials. If they request or demand a password, it is your right to refuse to comply, but equipment can still be detained for weeks or months. If this happens, you should consider legal assistance. —Ryan Lackey, Founder, Reset Security

ZERO-DAY Security researchers seek out vulnerabilities in code. When they find one, they have several courses of action. If they work for a government spy agency or a criminal gang, they may choose to create code that can exploit the vulnerability they have found—this weaponized code, before it is disclosed to anyone else, is called a "zero-day." It comes from the amount of time, in days, once the vulnerability is known until the maker of the software can fix the problem. On day zero (which is actually the first day—as computers always count everything starting from zero), the weapon is active. The ethics of selling zero-days is debatable. Companies that sell them to governments argue that, so long as the transaction is legal, the ethics are beside the point. Critics say that governments can use zero-days to attack and monitor dissidents. It's a tough call.

INFRASTRUCTURE ATTACKS In March of 2007, researchers at Idaho National Laboratory sent a test cyberattack to breakers that protected a 2.25-megawatt diesel-powered generator. Within a minute, the generator, weighing tons, literally jumped in the air, began to smoke, and was destroyed. Official video of this attack—considered the first public demonstration of a successful cyberattack on critical infrastructure—was leaked to CNN. The "Aurora Vulnerability," as it was called, was shocking for its simplicity, and cyber security experts began pointing out that America's supervisory control and data acquisition (SCADA) networks and industrial control system (ICS) networks are aged, fragile, overwhelmingly small, and privately owned—so this problem is not something that the U.S. government can simply order fixed. Ultimately, if a local power department decides not to invest $3,000 USD in patch management, that's a private business decision that the government can't overrule, absent clear threat and a court order.

The media became fascinated by attacks on SCADA and ICS, seeing every shutdown as a potential hack. Several attacks on critical infrastructure have happened, and each has been denied vocally by some. In 2009, widespread power outages in Brazil were reportedly caused by hackers; experts reported that it was soot, not hackers. Senior U.S. officials countered "nuh-uh," and it's never been settled.

Russian Aggressions No such uncertainty exists when it comes to Russian tactics: Russian government-mounted cyber attacks in the form of website takedowns, DNS attacks, and ultimately the complete blackout of Georgian internet traffic, which served as a precursor to invasion in 2008. This tactic has become a standard by Russia, which rather openly cyber-attacked the Ukrainian power grid in 2016, shutting down more than fifty power substations.

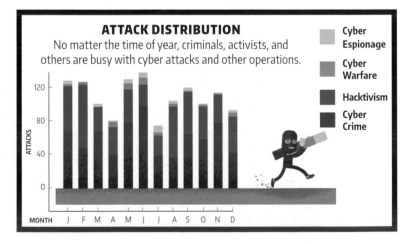

ATTACK DISTRIBUTION
No matter the time of year, criminals, activists, and others are busy with cyber attacks and other operations.

Cyber Espionage
Cyber Warfare
Hacktivism
Cyber Crime

ATTACKS

120

80

40

0

MONTH J F M A M J J A S O N D

What's at Risk? The scary news is that the SCADA systems in control of the nation's power are no any worse off than the systems that protect water, sewage, or other critical infrastructure, such as oil and gas. The good news is that, over the past few years, the federal government, along with the North American Electric Reliability Corporation and other groups, has been focusing intensely on SCADA and ICS issues. The problems are not yet solved, but we are in marginally more aware than we were a few years ago. That said, attacks have indeed been weaponized, and more things are connected to the internet than ever before (even though they shouldn't be), so it all may be a wash.

GOOD TO KNOW

PAGING YOUR INFRASTRUCTURE Does anyone use pagers anymore? Funny thing—apparently, a large number of critical infrastructure players, including chemical manufacturers, nuclear and electric plants, defense contractors, and others rely on unsecured wireless pagers to automate their industrial control systems. According to a 2016 report, this practice opens them to malicious hacks and espionage. In the report, researchers from security firm Trend Micro collected more than fifty-four million pages during a four-month span using low-cost hardware. In some cases, the messages alerted recipients to unsafe conditions affecting mission-critical infrastructure as they were detected. According to the report, "These unencrypted pager messages are a valuable source of passive intelligence, the gathering of information that is unintentionally leaked by networked or connected organizations. . . . Taken together, threat actors can do heavy reconnaissance on targets by making sense of the acquired information through paging messages. Though we are not well versed with the terms and information used in some of the sectors in our research, we were able to determine what the pages mean, including how attackers would make use of them in an elaborate targeted attack or how industry competitors would take advantage of such information."

T/F

AMERICA'S NUCLEAR LAUNCH CODES ARE KEPT ON FLOPPIES

TRUE Currently, if you want to launch a nuclear strike, you need technology from the 1970s in order to do so—an 8-inch floppy disk of the kind most modern adults have never even seen. It's not unreasonable to assume that this is on purpose—sort of a *Battlestar Galactica* scenario, where our most sensitive data is kept in such a way that our enemies (or teenage hackers) can't get at it. In fact, the truth is much more mundane—the U.S. government's cyber systems are woefully out of date, and we spend some $60 billion USD annually on maintaining those outdated systems.

MOBILE PHONE HACKS From sometime before 2004 (during the run up to the Olympic games in Greece) until January of 2005, either a criminal-based or nation-state-sponsored hacking gang engaged in a mass mobile phone network hacking operation. This group was tapping into Vodafone's switches in Greece, and targeting a range of calls to specific phones. Targets of this hacking incident included the mobile phones belonging to ministers of Greek national defense, foreign affairs, and justice departments; the mayor of Athens; the European Union commissioner for Greece; a variety of individuals engaged in civil rights, anti-globalization, and peace activism; and diplomats carrying mobile phones belonging to the U.S. embassy at Athens. Since that time, security experts have pointed out similar cases around the world in which the mobile networks in various countries have been compromised.

A History of Vulnerability None of these issues are particularly new—mobile networks, just like any other network, are susceptible to security issues. Nonetheless, occasionally this sort of thing becomes big news. In 2016, we learned of vulnerabilities in a digital signaling protocol that mobile phone carriers such as AT&T, T-Mobile, and Sprint use to track their users' identity, location, and more. A *60 Minutes* segment highlighted a hacker using that vulnerability to access phone data. That episode scared the pants off of seemingly everyone who saw it—and a bunch who just heard about it second-hand. Mobile networks are being patched as fast as they can, but SS7 (Signaling System No. 7, a common network protocol for handling phone calls and SMS messages) is only one of many vulnerabilities in how mobile operators handle the complex task of routing more and more voice and data calls each day.

MOBILE PRIVACY
Your mobile device's signal and data could be intercepted mid-transmission, and you might never know it.

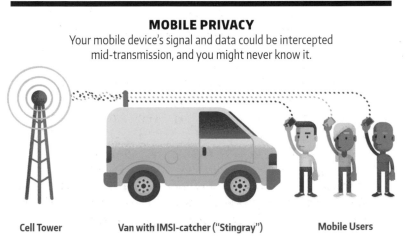

Cell Tower   Van with IMSI-catcher ("Stingray")   Mobile Users

> IF YOU TRAVEL ABROAD, FAMILIARIZE YOURSELF WITH LOCAL CYBER SECURITY LAWS, AVOID PUBLIC WI-FI IF AT ALL POSSIBLE, AND PROTECT YOUR DEVICES TO THE FULLEST EXTENT THAT YOU CAN, PHYSICALLY AND DIGITALLY.

Mobile Privacy Today During the past handful of years, the privacy community has begun to seriously question how good law enforcement really is at intercepting cellular signals and harvesting mobile phone data. The equipment for doing this sort of thing has been available to federal agencies and to some larger law enforcement agencies for several years by now. Technical advances have brought the costs down, while increased reliance on smartphones by individuals has increased the bang for the buck these products can provide, so more agencies are using them. These include tools called IMSI-catchers, which we've discussed briefly earlier in this book. IMSI stands for "international mobile subscriber identity"—that is, the unique identification number tagged to each mobile phone, which then allows a cellular network to distinguish each user from another. This device works as a man-in-the-middle platform for eavesdropping on phones on the GSM (global system for mobile) network.

Essentially, IMSI-catchers are portable base stations that can simulate a powerful cellular phone signal tower so that your phone, which always seeks out the most powerful signal within range, associates itself with it. Once that happens, the IMSI-catcher will intercept your signal before passing it on to a real tower (so that your call still does go through), but it captures everything that both sides say all the while—and you probably won't even notice.

Spying on the Airwaves Think IMSI-catchers are the thing you need to worry about if you want to avoid being eavesdropped on? Unfortunately, that's far from the case. The emergence of 4G LTE (long-term evolution) networking, also known as LTE, addressed some of these privacy issues, but, in 2015 researchers released information about kits that run about $1,200 USD and allow anyone who has a laptop and a universal radio software peripheral (USRP) and the proper software to intercept and locate 4G LTE traffic.

Another thing to consider is the advent of open platforms for mobile telephone operators. What Linux is to operating systems such as OpenBTS and OpenLTE are to mobile telephony: a set of freely available tools that can enable highly sophisticated mobile operations. David Burgess, one of the creators of OpenBTS, set up a mobile network at the Burning Man festival using his OpenBTS platform—and it worked just fine for everyone using it. As these tools to interact with increasingly smarter phones become less expensive and more commonly available, and as we rely more on our mobile devices for everything, we can expect even more attacks on cellular phones and mobile networks using this vector.

KILLER APP

BURN, BABY, BURN
A number of apps out there let you create a new, anonymous, and theoretically untraceable phone number that you can use from your mobile. These are helpful even if you're not engaging in international espionage. They're great for talking to potential dates, selling things on Craigslist, or in a dangerous situation where someone like an abusive spouse or parent is monitoring your calls. Here are some popular options:

Burner One of the best and easiest-to-use apps, but it only works in the United States and Canada.

Hushed Works in forty countries over VoIP, so it will cut into your data plan if you use your cellular internet connection.

CoverMe This app has numbers that appear to originate from the United States, Canada, UK, China, and Mexico.

ON A TRIP OUTSIDE THE COUNTRY, CONSIDER RENTING A LOCAL MOBILE PHONE OR COMPUTER TO AVOID ANY CUSTOMS ISSUES OR SEARCHING OF YOUR DEVICES UPON RETURN.

WHAT'S THE BUZZ? After 9/11, one of the most commonly heard terms from the intelligence community in the public domain was "chatter." It's a brilliant term because it sounds highly specific, but it's really quite generic, encompassing signals intelligence and publicly overheard sentiment on radio, television, internet forums and chatrooms, newspaper editorials, and gossip. Basically, chatter can mean just about anything.

We won't hear anything useful about SIGINT being gathered today (unless it's being proffered as a justification for military actions or economic sanctions), because our intelligence agencies focusing on that (mainly the National Security Agency) are incredibly good at not saying things. But we do understand from a few peeks inside how chatter is used by professionals to track terror groups.

Taking Responsibility If 9/11 resulted from a breakdown of intelligence upon a national scale, at the New York Police Department it was seen as a failure to take responsibility for the city's destiny. New York doesn't stay on the wrong side of a problem very long; at about 9 a.m. on September 11, 2001, senior officials at the NYPD surveyed the smoldering wreckage of the World Trade Center and said, "Yeah, this is not going to happen ever again." They began to create an intelligence capability that has become an extraordinary agency in its own right, combining tradecraft and procedures imported from agencies like the CIA (David Cohen, the NYPD's first commissioner for intelligence, is ex-CIA) and a hodgepodge of methods that other agencies practiced. The result was uniquely New York, because the NYPD got to do something few ever get to do: start from scratch and create a culture.

Doing It Right Like many other intelligence agencies, the NYPD created a cadre of incredibly capable young officers to be stationed around the world to provide continuous, on-the-ground intelligence. And one thing that the NYPD Intelligence Bureau had done very well was to establish it early on. Uniquely, for its cyber capability, the bureau used NYPD detectives to focus on looking for threats, capabilities, and intent, while also relying on civilian analysts to provide the language expertise (everything from Arabic to Pashto to Urdu) and for their cyber-fu. Meanwhile, the bureau also allowed the detective-investigators to remain professionals in their highly significant areas of expertise.

Typically, the IB Cyber Unit focuses on detection and investigation of radicalization and various threats as they pertain to New York City. In a nutshell, the team focuses on the enormous pile of people saying stuff that sounds radical, separating out people who are just spouting off or exercising free speech from those truly thinking about radicalization, then investigating and separating the curious from those with true intent.

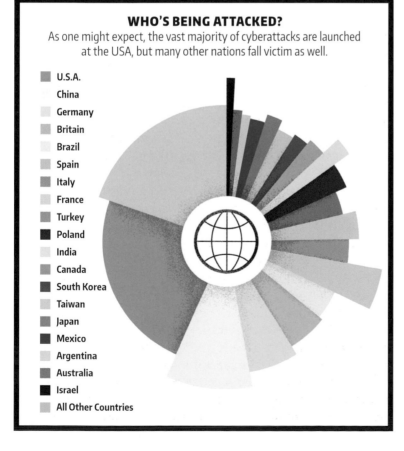

WHO'S BEING ATTACKED?

As one might expect, the vast majority of cyberattacks are launched at the USA, but many other nations fall victim as well.

- U.S.A.
- China
- Germany
- Britain
- Brazil
- Spain
- Italy
- France
- Turkey
- Poland
- India
- Canada
- South Korea
- Taiwan
- Japan
- Mexico
- Argentina
- Australia
- Israel
- All Other Countries

ONE SILVER LINING TO THINK OF REGARDING CYBER WAR ... AT LEAST IT'S EASIER TO REBOOT A NETWORK THAN REBUILD A CITY.

CYBER TERRORISM In the last five years, many of the cyber-attack tools that were once used exclusively by nation-states have become more easy to obtain, meaning that they can now also be used by criminal gangs and—at least in theory—terror groups as well. But buying a great piano really cheap doesn't mean you can suddenly play Chopin. The money and training that go in to a cyber operation is the true barrier to entry.

During the 2016 election, hacking by Russia caused tremendous disruption in the United States. We now know that during the six- to nine-month gestation period after the Russians gained entry to the network of the Democratic National Committee, but before they began to release email publicly, their activities consisted mainly of lateral movement within the network. During that time, the attackers engaged in rather routine but essential activities of a long-term network reconnaissance operation, including data classification and location. The hackers were answering the questions: What does the DNC have? Where do they keep it? How do they use it? How do they access it? Basically, they were learning the answer to "What does 'normal' look like in this organization?" All this showed one important difference between a nation-state attack and those mounted by terror groups: tradecraft.

Art and Craft Tradecraft is the techniques, methods, and tools that together form the art of spying, and it's not something that comes easily. It takes years of experience, lots of money, and great leadership and training. Mostly, when we look at terror groups, we see them spending what money and leadership and training resources they have not on tradecraft but on materiel and logistics for attack: moving men, guns, and bombs across distances; getting them training; smuggling them across borders; and mounting attacks.

Hackers Are Everywhere The barriers terrorists being able to launch a cyber attack are getting lower. When we look at the troubles that groups like Anonymous and LulzSec have caused law enforcement and other government groups, the disruption was significant. Their success was based on a commonly agreed-upon mission, a decentralized command and control, and the availability of free, easy-to-use, and easy-to-learn hacking and attack tools. This sounds like the basis of a classic terrorist attack, and it can be used by groups such as ISIS once the cyberweaponry they would need has be simplified to the point that it's easily adopted by groups with minimal resources. It just takes a small group of radicalized, computer-literate believers to tip these scales.

THE TAKEAWAY

Protecting your data when you travel is fairly easy. Stopping a global cyber war—not so much. Still, there are always ways to be prepared.

BASIC SECURITY
- Protect your IP online and when traveling.
- Encrypt all products and IP-related communications.

ADVANCED MEASURES
- Use purpose-built devices for cross-border travel.
- Maintain minimal mobile mail settings (no one needs more than thirty days of email on their phone at this point).
- Encrypt everything.
- Minimize data sets provided to business partners.
- Audit partners' security as you would your own.

TINFOIL-HAT BRIGADE
- Prepare for an infrastructure attack.
- Get off the electric grid with solar power.
- Prepare to have an interruption in your water supply.
- Practice self-sufficiency.

SHALL WE PLAY A GAME? In 1983, a movie called *WarGames* captured the world's imagination with the plotline in which a teenage hacker almost starts World War III when he hacks into a Defense Department computer. One major unexpected consequence of that film was that then-president Ronald Reagan saw it and was subsequently moved to devote resources into what we now call cyber security. A great number of young hackers were also inspired by the movie, perhaps most the famous being a group of Milwaukee-area high-school students who went by the collective name The 414s (for their area code in Wisconsin). The 414s breached a number of big corporate networks for fun, simply as a prank rather than to steal or damage anything. Since hacking was more difficult to prosecute at the time, they were finally charged with making illegal phone calls.

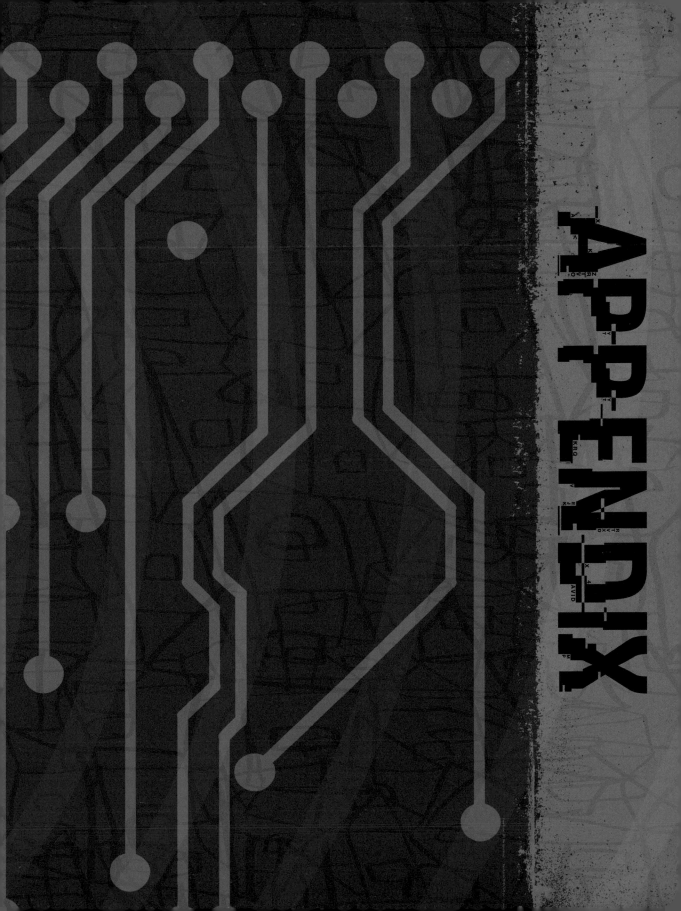

APPENDIX

THE BIG TAKEAWAY

Throughout this book, we've ended each chapter with a summary of what we've called "the takeaways"—some hands-on action items you can apply in your everyday life to avoid the dangers highlighted in that chapter. Avid readers have no doubt noticed that some of these measures are repeated multiple times, because some safety measures are widely applicable. Some other suggestions may only show up once, if they pertain to a very specialized area (such as online education, or international affairs). This chart is the "if you remember nothing else, remember these things" summary to clip out and hang on that smart, Wi-Fi-enabled refrigerator that for all you know is sharing scandalous tidbits about your snacking habits with foreign spies right now.

BASIC SECURITY

- Use different strong passwords for every login (website, desktop programs, phone apps).
- Use a password vault program.
- Password-protect and disable remote management on your modem, router, and any other Internet-connected devices using unique passwords.
- Password-protect home Wi-Fi and encrypt with WPA-2 PSK at a minimum—never WEP.
- Never share your login information with anyone.
- Don't click on suspicious links or download unexpected files.
- If anything you're offered online seems too good to be true, it is.

- Never give private information out over email or text. Always call the bank, utility, or service that's ostensibly asking for your information.
- If you lose your wallet, report missing cards immediately. Carry the minimum set of cards, and never your Social Security card.
- Set all social media privacy settings as high (private) as possible.
- Monitor kids' social media usage, and talk to them about online sharing and safety.
- Use a minimum of 8-digit screen lock codes (not fingerprint or face recognition) on all mobile devices.
- Encrypt your phone.

ADVANCED MEASURES

- Always use two-factor authentication (2FA) when possible.
- Don't get your children Social Security cards if possible.
- Check your credit report regularly; do so for all family members including kids.
- File a police report after fraud of any amount.
- Only use CHIP-and-signature cards (or CHIP+PIN where available).
- Only use the internet in incognito mode.
- Encrypt everything.
- Never use public Wi-Fi without a VPN or SSH tunnel.
- Restrict and lock down your home network, starting with DNS.
- Install GPS tracking apps on kids' phones.
- Limit location services and Wi-Fi use on your phone.
- Ensure the minimum metadata is saved with all photos.
- Only use credit cards that offer fraud and identity protection.
- Maintain minimal mobile mail settings.

TINFOIL HAT BRIGADE

- Eschew electronic communication wherever possible.
- File your taxes the old-fashioned way: on paper.
- Don't use banking apps on your phone.
- Don't shop online except through guest accounts and one-time credit cards.
- Don't shop at stores with older, swipe-only (non-Chip) POS terminals.
- Post online only under anonymous usernames; change them frequently.
- Lock down all social media accounts to private; ensure your children have done the same.
- Cover all computer webcams and microphones with electrical tape; remove cameras and microphones from mobile devices if you can.
- Use spyware to track all of your children's online activity.
- Use a private LAN for kids' computers, IoT devices, and TVs, and aggressively blacklist sites at the router.
- Use encrypted DNS.
- Regularly reflash your phone to factory settings.
- Prepare for an infrastructure attack with off-the-grid self-sufficiency measures.

WHAT'S NEXT

IF THERE'S ONE THING WE HOPE YOU GET OUT OF THIS BOOK (BESIDES NOT TO USE "PASSWORD" AS YOUR PASSWORD), IT'S THAT NEW THREATS AND NEW TOOLS ARE ALWAYS POPPING UP, BUT WITH COMMON SENSE, A BASIC KNOWLEDGE OF CYBER SECURITY, AND A HEALTHY DOSE OF SKEPTICISM YOU'LL BE IN GOOD SHAPE NO MATTER WHAT THE FUTURE HOLDS. HERE'S WHAT WE'RE THINKING ABOUT.

NICK SELBY

We are approaching an inflection point, at which consumers begin to demand security as a fundamental consumer right. The next five years will be tumultuous, as companies and governments worldwide test the public's commitment to this new realty: first, of course, with lip service and waving hands. I am optimistic about the future, though. More people are encrypting their communications every day, and applications to he,p them do so are finally becoming user friendly.

THE NEXT BIG THING(S) Political upheaval and cyber activism will combine in a storm of new defections by government employees and contractors releasing more code and program and strategy depictions. Foundationally insecure municipal, county, and state systems, as well as critical infrastructure, will be betrayed by attempts to provide app-based access-convenience to an IT fabric incapable of supporting it.

WHAT I'M EXCITED ABOUT The disruption of transportation industries on Earth and in space, along with new autonomous and energy technologies, will create opportunities while providing more data than ever conceived about how we live, travel, and interact.

THINGS THAT WORRY ME It still seems cheaper to build fast, get to market, and fix the bugs later. Several generations of medical technology—especially implantables—out there now were built that way, and vendors have shown they won't fix problems unless forced to. Until manufacturers truly adopt the idea that it's cheaper and better to fix security during development, the speed of innovation will result in unsound and dangerous products.

THINGS THAT DON'T Nation-states like Russia, China, France, and the United States hack. It's how the world works. Intelligence services conduct intelligence operations. It's their job, and it's necessary. Yelling about it won't help.

The world changed as we wrote this book. Hacking attacks as part of information operations against our government, political institutions, and businesses didn't just become mainstream knowledge, they became political footballs. Bad security exposed not just credit cards, but the deepest secrets of the most powerful people and countries on Earth. All that stands between us and better cyber security is customers refusing to accept insecure code or apps. Vote with your money. Support secure applications.

HEATHER VESCENT

As soon as people hear that I'm a futurist, they almost always ask for a few predictions. I'll let you in on a little secret: No serious futurist makes predictions. The best thing about the future? It hasn't happened yet. If you don't like the present, you can actively work to improve the future.

Although I won't make predictions, I certainly hope something in this book changes your future. Maybe you'll beef up your security settings or pause for a moment before blindly accepting cookies from that sketchy website.

THE NEXT BIG THING(S) Get used to hacks and security breaches, because they aren't going away. The silver lining to these corporate freakouts? Security will improve and software will get better for everyone. Hackers will adapt and new breaches will happen on all-new technologies (especially watch IoT).

WHAT I'M EXCITED ABOUT The Internet of dogs! Think, IoT + working dogs + dog-computer interfaces. Augmenting man's best friend makes him even more powerful. Search and rescue dogs can work on a much higher level simply by adding sensors to their harnesses. They can be taught to interface with their wearable technology, interacting with humans in a variety of environments. It might not be long before your dog can really engage you in conversation!

THINGS THAT WORRY ME Who owns the data we create every moment we spend online? Right now, it's not us. As we continue to augment ourselves with technology and create online personas in walled gardens, we may forfeit ownership of our online identities.

THINGS THAT DON'T Many of today's problems seem new and insurmountable, but we will solve these problems just like we solved the problems that came before them. For example: Global cyberwar is a big problem today, but like all

seemingly impossible situations, we'll solve it. We're living in the best and most exciting times, even if the natural byproduct of our innovations is a series of new problems.

Nostalgia is counterproductive and depressing. We can never go back to the way the world was, and that's largely a good thing. Learn what you can today to improve your tomorrow. Do your best to protect yourself and your family. Say sorry when you screw up. Accept the apologies of those who screw up with you and don't hold grudges. Working together, real effort yields real progress. I like to keep Richard Feynman's quote in mind: "We are at the very beginning of time for the human race. It is not unreasonable that we grapple with problems. But there are tens of thousands of years in the future. Our responsibility is to do what we can, learn what we can, improve the solutions, and pass them on."

GLOSSARY

BITCOIN: The first of the difficult-to-trace digital currencies. It uses cryptographic techniques for peer-to-peer transactions that are not linked to traditional banking institutions, allowing users to keep their identities separate from their online wallets.

BACKDOOR: A non-obvious entryway built into a computer system or software program for remote administration; also refers to a method of bypassing authentication security to enable an unauthorized person to gain access.

BIOMETRIC DATA: Mathematical representations of measurable physical characteristics, such as irises, fingerprints, facial structure, voice, etc., utilized to verify identification.

BLACK HAT HACKER: A hacker who breaks into systems and networks for malicious intent. The black hat hacker might steal data, install malicious code, or otherwise exploit the penetration.

BLOCKCHAIN: The general technology concept of a decentralized public ledger that a cryptocurrency uses to record all transactions.

BURNER PHONE: A disposable phone, bought for cash, to avoid tying identity to a number. Related, a burner number is a temporary phone number generated by an app that lets users keep their main number private, adding an extra layer of privacy to smartphones.

CAMMING: The use of a webcam to communicate with someone through the Internet, usually refers to sexual performance done for fun or money.

CATPHISHING: The practice of creating a false online profile for the purpose of deceiving people who are looking for genuine relationships.

CRYPTOME: Created in 1996 by privacy advocate John Young, this site collects information about freedom of speech, privacy, cryptography, dual-use technologies, national security, intelligence, and government surveillance.

CRYPTOCURRENCY: Operating independently of a central bank or much government regulation, cryptocurrency is a medium of exchange that uses encryption techniques to transfer funds anonymously and to regulate the creation of "new" currency.

CRYPTOGRAPHY: The enciphering and deciphering of messages as well as the storing and transmission of data in such a way that only those for whom the data is intended can read and process it.

CYBERCRIME: Generally, crime involving a computer or network; specifically, illegal activity committed through or leveraging an electronic-based medium or targeting a computer-based platform.

CYBER ESPIONAGE: Illicitly obtaining confidential or classified information from individuals, governments, or organizations, using malicious software, such as Trojan horses and spyware, or cracking techniques, for political, economic, or military advantage.

CYBERSTALKER: Harassing or frightening someone through electronic communications; referred to as "cyberbullying" when it involves children.

CYPHERPUNK: An individual who believes that personal privacy is imperative for social and political change advocates for strong cryptography and privacy-enhancing technologies to allow online anonymity.

CYBERWARFARE: The use of computer technology to attack or sabotage vulnerable information systems of a state or a nation for strategic or military purposes. Attacks encompass introducing viruses, disabling websites and networks, denying or disrupting essential services, stealing or altering classified data, among other nefarious possibilities.

DARKNET: An intentionally hidden, small segment of the deep web that is only accessible through special software and cannot be reached through standard web browsing techniques. Encryption and anonymizing software hide users' movements. The darknet is often used for accessing illicit content and illegal peer-to-peer file sharing, although whistle-blowers and political dissidents, among others, have valid reasons for utilizing it.

DEEP WEB: Making up 90 percent of the Internet, the deep web is essentially anything a search engine can't find, consisting of vast amounts of unindexed information. Examples include anything that is password protected, I.R.S. and Social Security information, members-only databases or emails, bank statements, etc.

DIGITAL NATIVES: The generation of kids born after the mass adoption of the Internet.

DISTRIBUTED DENIAL OF SERVICE (DDOS): A cyberattack whereby an online service is rendered unavailable by overwhelming it with traffic from multiple sources.

DOXXING: The practice of researching and broadcasting online private or identifiable information (especially personal information such as a home address) about an individual or organization.

DUMP: A credit card dump is when cyber thieves copy the information in the magnetic strip of an active credit card to make a fake credit card that can be used by cybercriminals to make purchases.

EXIF DATA: Short for Exchangeable Image File, this is the data your smartphone or digital camera collects when you take a photo, which, in addition to camera settings, can reveal when and where your image was taken, enabling stalkers to track you.

FIREWALL: The first line of defense in network security, firewalls monitor incoming and outgoing computer traffic and, based on predefined security rules, determine which traffic should be allowed or blocked.

FRAUDULENT DATA FURNISHING: Done by a complicit business insider, this scheme involves reporting fake financial data for a synthetic ID to credit bureaus. In more sophisticated schemes, the business doing the reporting may also be fictitious.

GRAY MARKET: The trade of legal non-counterfeit goods that are sold outside of the manufacturer's chosen distribution channels; gray-market goods, also known as "parallel imports," are usually much cheaper when purchased this way, but do not come with warranties.

HACKTIVISM: A portmanteau of "hacking" and "activism," hacktivism is the act of breaking into a computer system to promote social or political change.

HASH: Used to create signatures to authenticate messages and files, a hash is an algorithm that turns the contents of a file (such as text or an image) into a fixed-length value called a "hash value" or "hash code."

KNOWLEDGE-BASED AUTHENTICATION (KBA): A method some websites use for account verification. With static KBA, also known as "shared secret questions," users answer previously defined questions; with dynamic KBA, questions are generated based on what the site knows about the visitor, culled from public and private data. An example might be, "What was your street address when you were ten years old?"

KOMPROMAT: A portmanteau of the Russian words for "compromising" and "information", and a mainstay of espionage, this term refers to compromising materials about a politician or other public figure used to create negative publicity, or to use for blackmail or for ensuring loyalty.

EXPLOIT: An attack on a computer system, taking advantage of a vulnerability that allows unauthorized access.

IMSI-CATCHER: An eavesdropping device that impersonates cell-towers and intercepts cell phone traffic, tracking the movement of the user.

INFOSEC: The term refers to "information security," or the protection of information (both electronic and physical) from unauthorized access, use, destruction, etc.

INTERNET OF THINGS (IOT): A network of interconnected objects that send and receive data via the Internet using bluetooth or Wi-Fi.

'LEET (1337) SPEAK: An informal language used by hackers to conceal their sites from search engines that usually consists of replacing letters with numbers or symbols.

LULZ: Derived from the plural of lol (laughing out loud), this term primarily means "laughs" or "laughing," but the Internet term has come to mean laughing at someone else's expense.

MALWARE: Malware stands for "malicious software," programs that can damage or compromise those who unwittingly download it.

MMOG: A massively multiplayer online game.

MT. GOX: A Tokyo-based bitcoin exchange, which was launched in 2010 and filed for bankruptcy in 2014 after hackers apparently stole the equivalent of $460 million from its online "virtual bank." The origin of the names comes from Magic: The Gathering Online eXchange, the site's original use (exchanging Magic: The Gathering cards.)

such as a bank, designed to lure unsuspecting victims into giving out personal information online that will compromise their security.

PHREAKING: The act of illegally breaking into telecommunications systems (hacking), especially to obtain free calls.

RANSOMWARE: malicious software designed to encrypt files and backups, holding them hostage until the victim pays for the decryption key. Often uses the same techniques as phishing in order to trick the recipient into opening the attached program that initiates the encryption process.

RED MARKET: Economic activities banned by the state, such as drug dealing, arms trade, human trafficking, and the buying and selling of human flesh such as blood, bones, and organs.

SHODAN: A search engine that indexes and points to Internet-connected devices; Shodan's crawlers search the Internet seeking connected servers, webcams, printers, routers, and other devices.

SIGNALS INTELLIGENCE (SIGINT): Intelligence gathering by intercepting electronic signals and communications. This can include all forms of video, voice and data communications.

OAUTH: Shorthand for "open standard for authorization," which enables Internet users to authorize websites or apps to access their information on other websites without sharing their passwords.

OPENBTS: Open-source software that replaces much of the traditional infrastructure used in conventional cell phone networks.

PASSIVE DATA COLLECTION: Consumer data, including capturing user preferences and usage behavior, is gathered without actively notifying the user. The best-known example is websites that use browser fingerprinting techniques, including the reading of cookies and other trackers that track users from site to site.

PENETRATION TESTING (PEN TESTING): The practice of testing a computer system, network, or web application in order to find any vulnerabilities that an attacker could potentially exploit.

PHISHING: Messages, often email, posing as legitimate businesses communication from a known entity,

SILK ROAD: A darknet market, described as "the eBay for drugs" (and other contraband and nefarious services), it was started by Ross Ulbricht in 2011 before being shut down in 2014. The market used TOR and bitcoin to avoid detection. The site's buyer feedback system lent a sense of security to illegal transactions, and some merchandise offered was considered to be of high quality. Ulbricht was convicted of money laundering, computer hacking, and conspiracy to traffic narcotics and sentenced to life without parole.

SKIMMING: The practice of collecting data from the magnetic strip on a credit card (or debit card) with a camouflaged counterfeit card reader, usually affixed over the card slot on Automated Teller Machines, gas pumps and other point of sale systems. The information harvested is then sold, and ultimately transferred onto a blank card.

SOCIAL ENGINEERING: The use of deceptive means to manipulate individuals into revealing all kinds of personal information. Similar to confidence games.

SPEARPHISHING: Unlike phishing, which involves mass-emailing, spearphishing generally targets users within a single organization and appears to be from an individual or business known to the recipient.

STEGANOGRAPHY: From the Greek word meaning "covered writing," this refers to the practice of concealing messages within other seemingly innocuous messages, often used to supplement encryption.

SWATTING: Making a prank call to emergency services in an attempt to bring about the dispatch of a large number of armed police officers to a particular address.

SURFACE WEB: Anything that can be indexed by a typical search engine like Google.

SURVEILLANCE MARKETING: The practice whereby companies observe and exploit the information you generate while using their service.

SYNTHETIC IDENTITY: An identity created from real and fabricated information, which is then used to establish a credit profile and secure an easy-to-get, low-limit credit card.

TOR: An acronym for "The Onion Router," Tor is a protocol that encrypts data and sends it through a network of volunteer relays set up around the world, creating layers that conceal users' source IP address.

The anonymous network, which was originally developed by the U.S. Department of Defense, can only be accessed with special software and a properly configured web browser.

TRUECRYPT: Shut down by its developers, TrueCrypt was once the go-to full- and partial disk encryption software, creating what amounts to a cryptographically hidden, password-protected section of your hard disk.

TWO-FACTOR AUTHENTICATION (2FA): Also known as two-step verification, this is a method of verifying login identity by utilizing two (or more) authentication factors. There are three types of authentication factors: something you know (such as a password, PIN, or user name); something you have (such as a bank card or a one-time password token); and the something you are (typically a physical characteristic such as a fingerprint

or retina sample). It is possible to have multiple element but single-factor authentication a good example is a website that uses a password (something you know) and a knowledge-based authentication question (something you know).

TUMBLERS: Programs or sites that mix identifiable or "tainted" cryptocurrency funds with others to obscure the fund's original source. The tumbler can also rapidly mingle

fractions of bitcoins in multiple transactions, after which the bitcoins come out squeaky clean.

USER-MANAGED ACCESS (UMA): This OAuth-based access management protocol standard defines how developers enable selective secure data sharing of a smart object. UMA removes the security burden from a manufacturer and gives the consumer/owner power over their own data.

VIRTUALBOX: A software virtualization package that installs on an operating system as an application, enabling you to run multiple operating systems, inside multiple virtual machines, simultaneously.

VIRTUAL MACHINE (VM): An emulation of a computer system. A virtual machine is an operating system installed atop software that imitates the presence of hardware, providing the same functionality of a physical computer. The end user has the same experience as they would have on their dedicated hardware.

VIRTUAL SESSION: A period of use on a computer, when a hypervisor (a piece of software running on top of either an operating system or bare metal) fools the operating system into thinking it is running on hardware, when in fact it is running in memory.

VOICE OVER INTERNET PROTOCOL (VOIP): Hardware and software that enables phone service over the Internet.

VIRTUAL PRIVATE NETWORK (VPN): An encrypted connection between a user's computer resident in an untrusted Internet location and a point behind the firewall and within a private network. A VPN allows users to safely traverse public networks (such as a guest Wi-Fi hotspot) by protecting all communications within an encrypted "tunnel." This is termed virtual because the encrypted tunnel mimics the connectivity provided by an actual network a user would access via, for example, a dedicated fiber connection.

VULNERABILITY: A weakness in a software application, computer system, or a network that can be taken advantage of by attackers.

WARDRIVING: Also called "access point mapping," this is the act of searching for Wi-Fi wireless networks by a person in a moving vehicle, using a portable computer or smartphone. Also used to describe the act of gaining access to Wi-Fi networks that are unprotected or poorly protected.

WAREZ: Pirated, or illegally copied, software that is distributed through the Internet; offering warez is illegal in the United States, as it is considered a form of copyright infringement.

WHITE HAT HACKER: This term refers to an ethical computer hacker or a computer security expert who breaks into protected systems and networks to test and asses their security before malicious (or black hat) hackers can detect and exploit them.

ZERO-DAY OR 0-DAY EXPLOIT: An action that takes advantage of a software vulnerability that is previously unknown to the software's author. "Day zero" is the first day in the count of days between the time the vulnerability becomes known until the day it is patched. Security researchers who create 0-days can, depending on their motivations, collaborate with the software publisher to coordinate a patch and then announce, or, commonly in the criminal or nation-state world, never report the vulnerability and simply use or stockpile the exploit until such a time as it is considered appropriate to use it. There is a small but lucrative legal and illegal global industry in creating, trading, and selling exploits.

INDEX

ABOUT THE AUTHORS

NICK SELBY is a Texas police detective who investigates computer crime, fraud, and child exploitation. He consults law enforcement agencies on cyber intelligence and investigations. He is also a cyber-security incident responder with Secure Ideas Response Team. Nick founded the information security practice at industry analyst firm 451 Research. As an information security industry analyst, Nick conducted technical interviews with more than 1,000 start-ups and established security and intelligence firms, wrote hundreds of strategic briefs, advised U.S. and European government agencies on security and intelligence, and was regularly quoted as a subject-matter expert in the world's leading business and industry newspapers.

A frequent contributor to newspapers, including the *Washington Post* and *New York Times*, he was coauthor of *Blackhatonomics: Understanding the Economics of Cybercrime*; and technical editor of *Investigating Internet Crimes: An Introduction to Solving Crimes in Cyberspace*. He also coauthored *In Context: Understanding Police Killings of Unarmed Civilians*.

HEATHER VESCENT is a futurist, primarily known as an expert in cyber economics and cryptocurrency. Her far-reaching research, published by the *New York Times*, *CNN*, *American Banker*, CNBC, Fox, and the *Atlantic*, explores trends in the future of money, economies, identity, wearable tech, relationships, augmented intelligence, IoT, cybersecurity, and humanity.

Her talks have been featured at many conferences and events, including SXSW, TEDxZwolle, and Cyber Security Summit. She has been profiled and quoted everywhere from *Wired* to the *New York Times*, the *Atlantic*, *American Banker*, *Italian Elle*, and beyond. She has appeared as a tech expert on Fox News multiple times.

For more than a decade, her company, The Purple Tornado, has produced media that "visualizes the future in the present tense," including ten short films and documentaries about the future and more than thirty podcasts on subjects including money, wearable tech, and self-driving cars. Heather splits her time between Los Angeles and the Mojave Desert.

OTHER CONTRIBUTORS

ERIC OLSON works in the field of cybercrime and threat intelligence, including developing software and systems to combat phishing, identity theft, and cybercrime. He is currently vice president of intelligence operations at LookingGlass Cyber Solutions and volunteers for a nonprofit battling online child exploitation. Eric has a BA in Russian Studies from Hamilton College and an MBA from Georgetown University.

MOEED SIDDIQUI got his first IT in municipal government at the age of fifteen (two years before he became an Eagle Scout). While in college, he ran his own managed IT services business specializing in networking, security compliance for healthcare, government and financial services industries, and project management. He has worked with small businesses, Tier 4 data centers, and everything in between.

JOHN BEAR, PH.D. wrote his first book about distance education, *Bear's Guide to Earning Degrees by Distance Learning* in 1974, before the birth of the Internet. He spent a decade consulting for the FBI's Operation DipScam, helping to expose and shut down fraudulent schools and is the author, with retired FBI agent Allen Ezell, of *Diploma Mills: The Billion-Dollar Industry That Sold a Million Fake Degrees*.

AUTHOR ACKNOWLEDGEMENTS

NICK SELBY Grateful thanks to Mariah Bear, Ian Cannon, Jan Hughes, Allister Fein, Suzi Hutsell, and the rest of the Weldon Owen and Cameron teams for their incredible support, and to Rob James for cracking the whip. Special thanks to Ben Singleton and Moeed Siddiqui at SIRT. I am very happy to thank Aaron Barr, Daniel Clemens, Kevin Branzetti, Dave Marcus, Ryan Lackey, Rhett Greenhagen, Lance James, Allison Nixon, Mike Kearn, Will Gragido and others in the infosec space who gave so generously of their time and expertise. All the mistakes are mine. I also thank my co-authors, Heather Vescent, Eric Olson, Moeed Siddiqui, Amanda Nickerson, and John Bear.

HEATHER VESCENT Thanks to my parents, Pam Phillips, Rick Schlegel, and Donna Dee. To Ruth Waytz, Sarafina Rodrigeuz, and Rosie Pongracz, thank you for your support, friendship and love. For research insights, security tips, inspiration, and personal stories, thanks to Monica Anderson, Ian Danskin, Scott Froschauer, Ashish Gupta, Alex Kawas, Jennifer Ramsey, Rob Ryel, Tamara Struminger, Corwin Weskamp, and my anonymous dark web patrons. And thanks to horndogs and trolls on Reddit. You know who you are, but you probably don't know who I am. I spent many late nights reading way too many comment threads.

Many thanks to my coauthor Nick Selby, and to the staff at Weldon and beyond: Mariah Bear, Ian Cannon, Jan Hughes, and everyone else. Apologies to anyone I have inevitably left out.

This book couldn't have been written without my late-night companions, coffee from Peet's and wine from Ravenswood and Bonny Doon. Finally, thanks to Mr. Dog, who doesn't need to worry about cyber security for now, because the Internet of dogs is still a few years away.

ILLUSTRATION CREDITS

Eric Chow: Front cover, 1, 2, 4, 8, 14–15, 16, 18, 22, 24, 28, 34, 36, 40 (bottom), 42, 46, 50, 52, 53, 58, 60, 64, 68, 69, 70, 72, 74, 76, .84, 86, 88, 92, 94, 96, 98, 100–101, 102–103, 104, 110, 112, 116, 122, 123, 124, 126, 128, 130, 132, 134, 136, 150, 152, 155, 156, 158 (top), 160–161, 164–165,174,178, 180, 182, 186, 188, 196, 198, 200–201, 207, 209, 210,211, 212, 216, 218, 221, 222, 224

Conor Buckley: 20, 25, 26, 30–31, 32–33, 38, 39, 40 (top), 44, 49, 56, 57, 62, 66–67, 75, 78–79, 80, 81, 82, 90, 91, 106, 108, 114, 119, 120, 125, 133, 138, 140, 144–145, 146, 148, 154, 158 (bottom), 159, 160, 166, 169, 170, 172–173, 181, 190, 192, 194, 197

weldon**owen**

President & Publisher Roger Shaw
SVP, Sales & Marketing Amy Kaneko

Associate Publisher Mariah Bear
Project Editor Ian Cannon

Creative Director Kelly Booth
Art Director Allister Fein
Senior Production Designer
Rachel Lopez Metzger

Production Director Michelle Duggan

WELDON OWEN would like to thank Rob James, Marisa Solis, and Brittany Bogan for editorial assistance and Ken DellaPenta for the index.

Weldon Owen is a division of Bonnier Publishing USA

1045 Sansome Street, #100, San Francisco, CA 94111
www.weldonowen.com

ISBN-13: 978-1-68188-175-1

10 9 8 7 6 5 4 3 2 1
2017 2018 2019 2020 2021

Printed and bound in China.

Produced in conjunction with Cameron + Company

Publisher Chris Gruener
Creative Director Iain R. Morris
Senior Designer Suzi Hutsell
Junior Designer Rob Dolgaard
Managing Editor Jan Hughes

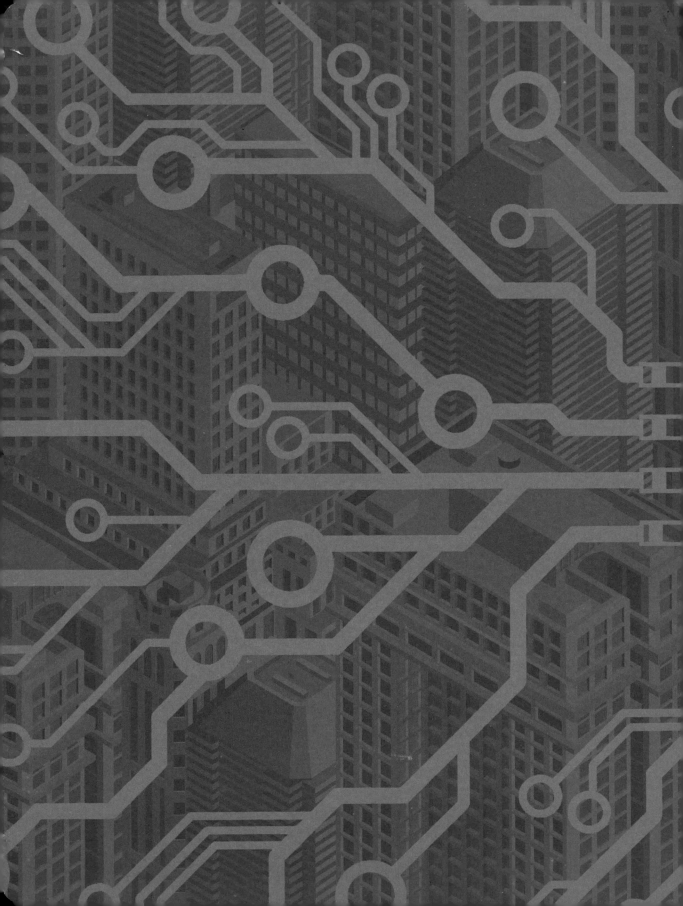